BISHOP OF EVERYWHERE

At present we are at a crisis in which one party is keeping the Bible in the clouds in the name of religion, and another is trying to get rid of it altogether in the name of Science. Both names are so recklessly taken in vain that the Bishop of Birmingham has just warned his flock that the scientific party is drawing nearer to Christ than the Church congregations. I, who am a sort of unofficial Bishop of Everywhere, have repeatedly warned the scientists that the Quakers are fundamentally far more scientific than the official biologists. In this confusion I venture to suggest that we neither leave the Bible in the clouds nor attempt the impossible task of suppressing it. Why not simply bring it down to the ground, and take it for what it really is?

Bernard Shaw, Postscript to
The Adventures of the Black Girl in Her Search for God

His plays are sermons – clever, rollicking, crazy, sincere, witty, brilliant and perverse – withal sermons. . . . He is a better bishop than a playwright.

J. F. Huneker, reviewing Three Plays for Puritans *in the (New York)* Musical Courier, *May 1901*

BISHOP OF EVERYWHERE

Bernard Shaw and the Life Force

Warren Sylvester Smith

The Pennsylvania State University Press
University Park and London

Library of Congress Cataloging in Publication Data

Smith, Warren Sylvester, 1912—
Bishop of Everywhere.

Includes bibliography and index.
1. Shaw, Bernard, 1856–1950—Religion and ethics. I. Title.
PR5368.R4S55 822′.912 81-17700
ISBN 0-271-00306-5 AACR2

Designed by Dolly Carr
Printed in the United States of America

For Paula, Rodney, and Selden . . .
and those who come after.

Contents

Acknowledgments

Portions of this book have appeared in *The Shaw Review; Shaw, the Annual of Bernard Shaw Studies; The Educational Theatre Journal; The Nation; Bulletin of the Friends Historical Society; The Pulpit;* and *The Teilhard Review.* They are used here by permission of the editors and publishers of those periodicals.

Quotations from the works of Bernard Shaw are by permission of the Society of Authors on behalf of the Estate of Bernard Shaw.

Acknowledgment is made to Sir John Maud and *The Times* of London for permission to quote from the letters of the Bishop of Kensington in chapter 4.

In chapter 16, the letters from Dame Laurentia McLachlan are used by permission of the Abbess of Stanbrook (February, 1978).

I am obliged to Stanley Weintraub for referring me to Shaw's unpublished lecture notes, discussed in chapter 3. I am especially indebted to John M. Pickering for his advice and encouragement in the preparation of this manuscript.

W.S.S

Prologue

Shaw vs. Shaw vs. Shaw

The play that most clearly gives us the three dominant compass-points of Bernard Shaw's mature personality is also the play that first brought him into prominence as a playwright in London — though it is seldom referred to these days and rarely revived. But in 1905 King Edward VII commanded a special performance of *John Bull's Other Island,* and in 1911, as part of the coronation festival for the new King George V and his Queen, Prime Minister Asquith arranged for a portion of the play to be done at No. 10 Downing Street.

The title was a lively one at the time, but it connotes little to the present playgoer. Instead of directing attention to the character (as in *Candida, Caesar and Cleopatra, Major Barbara,* and *Saint Joan*) or to the underlying theme (as in *Arms and the Man, The Devil's Disciple, Man and Superman*), it calls attention more than any other of Shaw's titles to a political relationship which no longer exists. Indeed we now have to remind ourselves that the title refers to Ireland. Perhaps it would have been better for survival purposes if Shaw had made the play a sort of companion piece to *Captain Brassbound's Conversion* and titled it *Tom Broadbent's Conquest.*

Shaw himself appears to have thought the play somewhat dated when he wrote the Preface to the Home Rule Edition in 1912. "Broadbent is no longer up to date," he declared. "The controversies about Tariff Reform, the Education and Licensing Bills, and the South African War, have given way to far more vital questions. . . . There is little left of the subjects that excited Broadbent in 1904 except Home Rule. And Home Rule is to be disposed of this year" (II, 875).[1]

The immediate issue of the play is one Englishman's conquest of a little Irish village, Rosscullen. Tom Broadbent, a civil engineer, acting on behalf of a syndicate, goes to Rosscullen with Larry Doyle, his Irish-born partner, to foreclose on some properties, and to turn the place into a garden city that will be a profitable tourist center. With irrepressible optimism and complete insensitivity except for a kind of political second-sense, he secures land-holdings, puts himself up to represent Rosscullen in Parliament, and gets himself engaged to the village's only "heiress."

That is the thread of the plot on which the play runs, but to reach its end it must travel through a maze of personalities, incidents, and topics. It is the topics that have become dated. But Shaw was somewhat premature in his 1912 estimate. Home Rule was not disposed of that year. There was to be much bloodshed, including the Easter Monday Rebellion of 1916 and England's savage suppression of it in the midst of World War I, before dominion status was granted to the Free State in 1922; and it was 1948 till the country became completely independent. A development Shaw had never imagined has kept "the Irish question" tragically alive — the refusal of the six Ulster counties to join with the predominantly Catholic Free State to the south. Shaw dealt with this later disillusionment in a 1929 Postscript to his original 1906 Preface for Politicians. He recalled that he had "exhorted the Protestants to take their chance, trust their grit, and play their part in a single parliament ruling an undivided Ireland. They did not take my advice" (II, 891). Consequently John Bull still has troubles with that other island, giving the play a revived historical interest. Even so, the topics Shaw listed in 1912 have, as he predicted, become meaningless to modern audiences; and there is little in the play that foreshadows the present troubles.

The play was originally conceived at the suggestion of W. B. Yeats for production at the struggling Abbey Theatre in Dublin. When the play was finished, the Abbey begged off, and it became instead the most successful Shaw play to that date in London. It is the only play that Shaw wrote specifically about his homeland, and even he, it seems, placed too much significance on that fact. In reality, except for a very few passages, the play is no more held back by topical references than Ibsen's *Ghosts* is by its topical (and mistaken) nineteenth-century references to syphilis.

Plays of stature — and *John Bull's Other Island* is certainly one — move on more than one level, the topical one often being the merest excuse for character revelation, philosophical rumination, or a heightened perception of the world. Consequently playwrights, like other artists, sometimes speak to the future in ways they could not have anticipated. How could Shakespeare have known what Shylock would mean to a twentieth-century audience, or how Hamlet would appear in a post-Freudian, post-Kierkegaard world? Shaw, closer to our own time, *might* have guessed what in effect has taken place — that Tom Broadbent has become an American. Already in 1906 he was observing:

> The successful Englishman of today, when he is not a transplanted

> Scotchman or Irishman, often turns out on investigation to be, if not an American, an Italian, or a Jew, at least to be depending on the brains, the nervous energy, and the freedom from romantic illusions (often called cynicism) of such foreigners for the management of his sources of income. At all events I am persuaded that a modern nation that is satisfied with Broadbent is in a dream. Much as I like him, I object to be governed by him, or entangled in his political destiny. [II, 810]

The man of means, swinging his influence abroad and taking the maximum advantage of his superior economic leverage, is no longer the Englishman — invading Ireland or any other part of the world. Consider Broadbent's "We'll take Ireland in hand, and by straight-forward business habits teach it efficiency and self-help on sound Liberal principles." In his impatient explanation to Keegan, too, Broadbent talks American: "The fact is, there are only two qualities in the world: efficiency and inefficiency, and only two sorts of people: the efficient and the inefficient" (II, 1013). No, Tom Broadbent is not really out-of-date.

America has failed as completely in Southeast Asia as ever Britain failed in southern Ireland (and is currently failing in Northern Ireland). Since World War II we have sent Broadbents and Doyles to all the world's islands of inefficiency, enlarging Broadbent's substantially political dream of salvation to a slightly more comprehensive socio-economic-political one. John Bull's Other Island has become Uncle Sam's Other Hemisphere.

Both the Broadbents and the Doyles are committed to the gospel of efficiency — the Broadbents with the ingenuous optimism of the salesman, the Doyles with the more painful scepticism of the scientist. Here and there, now and then, they are forced to confront the disturbing figure of Peter Keegan, the defrocked Irish priest, who knows that the salvation of this planet is feasible only through the spiritual regeneration of the persons living on it. The Keegans are a sensitive lot, and understandably bitter:

> This world, sir, is very clearly a place of torment and penance, a place where the fool flourishes and the good and wise are hated and persecuted, a place where men and women torture one another in the name of love; where children are scourged and enslaved in the name of parental duty and education; where the weak in body are poisoned and mutilated in the name of healing, and the weak in character are put to the horrible torture of imprisonment, not for hours but for years, in the name of justice. It is a place where the hardest toil is a welcome refuge from the horror and tedium of pleasure, and where charity and good works are done only for hire to ransom the souls of the spoiler and the sybarite. Now, sir, there is only one place of horror and torment known to my religion; and that place is hell. Therefore it is plain to me that this earth of ours must be hell, and that we are all here . . . to expiate crimes committed by us in a former existence. [II, 990–91]

Broadly this play pits the power of politics against the power of religion. And the theme, so far as it is possible to state such a thing in brief, must be the thoroughly timely one: the futility of political and economic reform without spiritual reform. For here and now, the politician-engineer wins, of course. Broadbent will marry his heiress, build his golf links, and win his seat in Parliament. But Keegan leaves no room for doubt as to the emptiness of the victory. "For four wicked centuries the world has dreamed this foolish dream of efficiency; and the end is not yet. But the end will come" (II, 1018).

This preoccupation with salvation — the bold name for all of this — is to be found not only in *John Bull's Other Island.* It runs like a broken thread through all the plays from *Man and Superman* (1903) onward. But in this one play more than in any other Shaw pits different *kinds* of salvation — different gospels — against one another; and in so doing he reveals three facets of his own character, providing the spectacle of Shaw vs. Shaw vs. Shaw.

Broadbent is described as "*a robust, full-blooded, energetic man in the prime of life, sometimes eager and credulous, sometimes shrewd and roguish, sometimes portentously solemn, sometimes jolly and impetuous, always buoyant and irresistible, mostly likeable, and enormously absurd in his most earnest moments*" (II, 894). At first it may seem impertinent to identify this man as a kind of Bernard Shaw, for it is at this social relative of Reverend Morell of *Candida* (and the later John Tarleton of *Misalliance*) that the most satiric laughter of the play is aimed; and we have become accustomed to satirists of lower stature who more cautiously place the butt of laughter safely beyond their own gates.

Much of Tom Broadbent is drawn from the youthful immature Shaw. Near the end of the play, Broadbent tells us of an early dream:

> Once, when I was a small kid, I dreamt I was in heaven. . . . It was a sort of pale blue satin place, with all the pious old ladies in our congregation sitting as if they were at a service; and there was some awful person in the study at the other side of the hall. I didn't enjoy it, you know. [II, 1021]

This is in fact a dream that Shaw himself had described at greater length ten years before, recalling the days when he was forced to go to church:

> I remember at that time dreaming one night that I was dead and had gone to heaven. The picture of heaven which the efforts of the then Established Church of Ireland had conveyed to my childish imagination was a waiting room with walls of pale sky-coloured tabbinet, and a pew-like bench running all round, except at one corner, where there was a door. I was, somehow, aware that God was in the next room, accessible through that door. . . . A grimly-handsome lady who usually sat in a corner seat near me

> in church, and whom I believed to be thoroughly conversant with the arrangements of the Almighty, was to introduce me presently into the next room — a moment which I was supposed to await with joy and enthusiasm. Really, of course, my heart sank like lead within me at the thought; for I felt that my feeble affectation of piety could not impose on Omniscience, and that one glance of that all-searching eye would discover that I had been allowed to come to heaven by mistake. Unfortunately for the interest of this narrative, I awoke, or wandered off into another dream, before the critical moment arrived.[2]

In politics, too, Broadbent is a thoroughly immature Shaw. It is his full-time delusion, reflecting Shaw's earlier part-time belief, that the world can be saved by political means. When Doyle is in a fit of despondency arising from anything but political sources, Broadbent cheers him with "Dont despair, Larry, old boy: things may look black, but there will be a great change after the next election" (II, 920). Even to Keegan's carefully assembled evidence that this world is in reality hell, Broadbent's observation is that "Of all the evils you describe, some are absolutely necessary for the preservation of society, and others are encouraged only when the Tories are in office." Indeed he can see "no evils in the world — except, of course, natural evils — that cannot be remedied by freedom, self-government, and English institutions" (II, 991).

These, it must be quickly noted, are not the remedies, precisely, that Shaw himself would have recommended! But Shaw did have his own bag of political remedies nevertheless. And when he was nearing fifty in 1904, he could even then look back at his years with the Webbs and the Fabian Society, his lectures and debates, and see in himself much of what is bitterly funny about Tom Broadbent. Though Shaw was always politically further to the left than Broadbent could ever venture, it would have been possible — even obvious — to place Broadbent all the way at the reactionary right as a dramatic element in the play. I think Shaw chose to depict a Liberal and "a bit of a Unitarian" for the barb of his satire because he could fathom the pompousness of a Liberal with more self-knowledge than he could give to the pompousness of a Tory and a churchman. It must be remembered that Shaw's earliest reputation in London was that of a Socialist orator and pamphleteer. The later playwright, surveying his own passion for reform and the degree to which he sometimes thought such reform could be achieved by political means, found the material to caricature the windbag, the carpet-bagger, the charlatan, and the fool — who believed himself all the while to be talking "straight common sense" and taking "his stand on the solid ground of principle and public duty" (II, 988).

Nothing could be more absurd than to carry the analogy too far. Broadbent *is* a caricature, an oversimplification and an amplification of one small part of Shaw's character. Indeed it is so small a segment that Broadbent is unable to see beyond the circle of his own pettiness (he considers his father mad because he joined the Tariff Reform League). There is enough Shaw that a portion of himself can be held up to satire; there would not be enough Broadbent. Certainly in his lack of control of sentiment, in his lack of humor, in his complacent optimism, Tom Broadbent is totally un-Shavian.

More recognizable to most of us is that portion of himself which Shaw dramatizes in Broadbent's partner, the Irish expatriate, Larry Doyle, "*a man of 36, with cold grey eyes, strained nose, fine fastidious lips, critical brows, clever head, rather refined and good-looking on the whole, but with a suggestion of thinskinnedness and dissatisfaction that contrasts strongly with Broadbent's eupeptic jollity*" (II, 901–02). Only an Irishman who had turned his back on Ireland could render the long, bitter, nostalgic passage in Act I in which Doyle wrings from his memory the poetry and the futility of the Irish character. This, the longest speech in the play, is an assignment in emotional memory and verbal imagery that will give pause to the best actor; but it is a worthy challenge, a pouring out of things too long buried in the heart and evoked with pain only by the imminent possibility of a return to the country of his boyhood. It is, according to a stage direction, "a passionate dream" and there can be little doubt that it is Shaw's own.

It is even likely that Shaw shared some of Larry's reluctance to return because of an earlier sentimental love affair. We know little of Shaw's first twenty years in Dublin, and Ben Rosset in his *Shaw of Dublin* makes a very plausible claim that Shaw purposely obscured much of his formative life from biographers. But a piece of youthful verse, scrawled in an old appointment book now in the British Museum,[3] seems to refer to a romantic attachment when Shaw was seventeen or younger, during the time when the Shaws were spending summers at Torca Cottage at Dalkey, a few miles out of Dublin, near the sea. The girl, poetically referred to as "Calypso," unlike Larry's Nora Reilly, married another. But such lines as "She proved a too perilous plaything" fit Larry's response to the girl who had joined their household when he was seventeen. To Broadbent's question, "Were you at all hard hit?" Doyle replies, "Not really. I had only two ideas at that time: first, to learn to do something; and then to get out of Ireland and have a chance of doing it. She didnt count. I was romantic about her, just as I was romantic about Byron's heroines or the old Round Tower of Rosscullen; but she didnt count any more than they did" (II, 918).

Shaw's official biographer, Archibald Henderson, merely mentions the teen-age affair,[4] but Shaw later told Stella Campbell of it, and it is probably referred to in *Man and Superman* when Tanner tells Ann about the "Rachel Rosetree" incident (II, 569–70).[5] But Shaw, in his late teens, was not wasting away in a little Irish village. He was living with his abandoned father in a dismal Dublin flat. Doubtless he had the same two ideas that

Larry had, and any minor romance "didnt count." Almost prophetically the last stanza of the jejune romantic poem breaks into self-mockery:

> Then farewell, oh heartbreaking Calypso
> Thou didst shape my philosophy well
> But believe me the next time I trip so
> No poem shall tell.[6]

This ability to shift into humorous detachment was to make Shaw a satirist. The conflict of Irish imagination with English facts simply drove Doyle into bitterness.

But the Doyle bitterness continues to underlie the Shaw satire. Just a few moments before, Doyle has been exclaiming, "Why man alive, look at me! You know the way I nag, and worry, and carp, and cavil, and disparage, and am never satisfied and never quiet, and try the patience of my best friends." And a little later he is to confess: "I wish I could find a country to live in where the facts were not brutal and the dreams not unreal" (II, 907, 919).

Brutal as the facts are, Larry Doyle does not run from their true meanings. He can classify Matt Haffigan's industry in "making a farm out of a patch of stones on the hillside" as the industry of a coral insect. He can relate Nora's delicacy to her diet. He can reject the offer of a parliamentary seat by telling the property owners that they are little men who have "neither honor, nor ability, nor capital, nor anything but mere brute labor and greed." He can say of Matt Haffigan's future, "I say let him die, and let us have no more of his like. . . . The real tragedy of Haffigan is the tragedy of his wasted youth, his stunted mind, his drudging over his clods and pigs until he has become a clod and a pig himself . . ." (II, 1014). He can be, that is, the Shaw who believes in facing the bitterest truth without the sugar-coating of sentiment and false optimism.

I think it might strike even the most casual member of the audience that Broadbent & Doyle, in addition to being a firm of civil engineers, are in other ways something of an entity. They are almost perfect foils for each other, each relying on the opposite qualities of his colleague to complete his own existence. At more than one place in the play the dialogue between them sounds like one man locked in his room, arguing with himself.

Neither of these two men, however, is an artist; and even taken together they do not add up to one. The artistic-religious facet of Shaw's character the playwright has saved to reveal in heavy concentration in the person of Peter Keegan, "*a man with the face of a young saint, yet with white hair and perhaps 50 years on his back . . . in a trance of intense melancholy, looking over the hills as if by mere intensity of gaze he could pierce the glories of the sunset and see into the streets of heaven*" (II, 922). Keegan had been defrocked because, by his own admission, he was unable to administer last rites to a dying Hindu, who had revealed the mystery of this world to him and had left him powerless in his priestly office. Like St. Francis, he treats all living things as his kin (we first meet him in conversation with a grasshopper). He gently scolds Nora for picking a flower: "If it was a pretty baby you wouldnt want to pull its head off and stick it in a vawse o water to look at." In the midst of the hysterical merriment occasioned by Broadbent's running over Haffigan's pig, Keegan is in deep remorse. Though held in awe by the villagers, he is generally thought to be a little mad. It is this gentle heretic who dreams Shaw's own dream of Heaven. In contrast to Broadbent's "sort of pale blue satin place,"

> In my dreams it is a country where the State is the Church and the Church the people: three in one and one in three. It is a commonwealth in which work is play and play is life: three in one and one in three. It is a temple in which the priest is the worshipper and the worshipper the worshipped: three in one and one in three. It is a godhead in which all life is human and all humanity divine: three in one and one in three. It is, in short, the dream of a madman. [II, 1021]

It is not surprising, of course, that Shaw has chosen a *defrocked* priest ("I am violently and arrogantly Protestant by family tradition," he writes in the Preface); the surprising thing may be that Keegan is in reality an *almost* orthodox Catholic, whose departures from the faith are certainly more technical than spiritual, and who would be regarded with as great respect by most living Catholics today as he apparently was in Ireland in 1904. While *John Bull's Other Island* was being written, another Irish priest, Father George Tyrrell, was having theological differences with his Church. A couple of years later (life following art) he, too, was defrocked and eventually excommunicated. Shaw may have known of Tyrrell's conflict with Pius X, though the storm over him had not yet broken publicly, and Tyrrell may have contributed to the image of Keegan. Tyrrell, too, was a gentle and quietly stubborn person. Like Keegan, he refused to allow Catholic orthodoxy to fetter his own thinking, but remained, in his own mind, at least, a devout Catholic, struggling to believe in the possibility of a heaven on earth to replace the hell that is here now. "Could you have told me this morning where hell is?" Keegan asks Larry. "Yet now you know it is here. Do not despair of finding heaven; it may be no further off." But Tyrrell, like Shaw, was a Dubliner, not a villager, and a far more outspoken reformer than Keegan. Keegan was content to dream his dream of madness, alone at the Round Tower. Tyrrell carried his disillusionment to the Vatican itself, with inevitable and tragic consequences.[7]

Raymond S. Nelson gives some good evidence that in addition to Father

Tyrrell, Shaw's friend Edward Carpenter may have contributed to the realization of Keegan.[8] But all such analogies must be limited. Peter Keegan is no more Father Tyrrell or Edward Carpenter than he is Bernard Shaw. Yet the cards are stacked in Keegan's favor at the end, and there can be little doubt of Shaw's sympathy with the madman's dream.

As soon as we are well acquainted with the trio of Broadbent, Doyle, and Keegan, we discover overtones of them in all of Shaw's mature works. Again and again we find ourselves confronted with the Broadbent-Shaw, the Doyle-Shaw, or the Keegan-Shaw. Only in this one play, however, are the personalities so clearly defined, so neatly opposed for dramatic conflict. Not until the final ten minutes of the play — after the issues of the play have been resolved, really — are the three left alone on stage together. Yet this brief clash of their gospels generates one of the most incandescent passages in the modern theatre.

Part I

The Emergence of the Life Force

1

The Temptations of Judas Iscariot

More than twenty-five years ago, at the time of the centennial of Shaw's birth, I published an article dealing with his religion. An actor friend who had himself played in a number of Shaw's plays expressed astonishment. "I had always assumed Shaw was an atheist," he said. Shaw's religious views had been mulled over in print long before I chose to write about them. G. K. Chesterton had given them sympathetic, though critical, attention in 1909; and since then every writer who has dealt seriously with Shaw has been forced to consider them. For C.E.M. Joad[1] and Eric Bentley[2] the religious Shaw was central. Still, the assumption of my actor friend was far from unusual.

Possibly for some readers and playgoers such a misapprehension still exists, but I think in the meantime it has become considerably easier to accept Shaw as "a sort of Unofficial Bishop of Everywhere," as he designated himself in his later years.[3] Perhaps this is because, as the actual image of the GBS-jokester-iconoclast has begun to fade into oblivion, the underlying message of his plays and prefaces is finally filtering through the entertainment. But more likely it is because, for the playgoing and preface-reading public, at least, the conception of religion itself has changed. Shaw has recounted that in his own youth he called himself an atheist "because belief in God then meant belief in a tribal idol called Jehovah; and I would not, by calling myself an Agnostic, pretend I did not know whether it existed or not."[4] Within most religious establishments the tribal idol against which Shaw and so many of his contemporaries rebelled has gone out of fashion, despite the clamorous efforts of some of the fundamentalist sects to reinstate him. Pope John XXIII, the Bishop of Woolwich, and Martin Buber (among others) have intervened in the meantime to make the orthodoxies generally less confining, so that church attendance need not necessarily be an intellectual embarrassment. Outside the churches and synagogues, too, the spirit of pluralism prevails in a variety of cults that seek the Universal Spirit with or without the aid of drugs or Oriental mysticism. Even during his own lifetime Shaw sensed this change, though he did not foresee the concomitant religious regression that threatens to polarize our society both religiously and politically. At the age of ninety-two he wrote:

> When I said many years ago that the Holy Ghost is the sole survivor of the Trinity, and that it is far more scientific to describe Man as the Temple of the Holy Ghost than as an automaton made of a few chemicals in which some carbon got mixed accidentally, I was accused of advertising myself by uttering paradoxes of the same order as the statement that black is white, which is not a paradox but a lie. Now that I am old and obsolescent, young people who happen to have heard about the paradoxical Shaw from their elders, and are tempted to read him, cannot find anything startling in me. If they have the requisite erudition, they point out that what I have said had been said long ago by St. Augustine and all the great spiritual leaders of mankind before and after him.[5]

Even so, the terms Shaw has specified for a viable religion are still essentially heretical, and he could still ask, "Where in the world is there a Church that will receive me on such terms, or into which I could honestly consent to be received?" (VII, 396).

Much of Shaw's value in any contemporary religious discussion derives from his unusual detachment. Just as he could never join a church, neither could he align himself with the Secularist movement of Charles Bradlaugh or with any of the other militantly atheistic Bible-smashers of the late nineteenth century. He never had a creed to defend or revolt from. A number of times he tells the story of his encounter with the phrenologist, who, presumably, could read a person's character through the shape of his or her skull.

> Before I had shewn my hand on the subject, he said: "I can see you are a sceptic." "How do you know?" said I. "I am a bit of a phrenologist," said he. "Oh," said I, "have I no bump of veneration?" "Bump of veneration!" he exclaimed: "Why, it's a hole." And as a matter of fact that part of the scalp which in very devout worshippers and very obedient moralists rises into a dome or a ridge like a church roof, is on my scalp a majestic plain with a slight depression in the centre.[6]

Whether or not the story is fictional, Shaw always regarded himself as the Complete Outsider, and felt that he was "a sojourner on this planet, not a native of it."[7] The world's institutions did not inspire him with awe.

Emotionally detached as he was, he always looked on the religious phenomenon with uncommon interest, perhaps by virtue of his having been "a bloody Protestant" in the midst of Catholic Dublin ("a *sanguinary* Protestant" is the way Puritan Shaw put it in public). On several occasions he reminisced on his religious upbringing — or rather on the lack of a religious upbringing, for the difference between Catholic and Protestant was an economic and political difference more than a religious one. In any case his parents were not deeply religious people, and, though he had been properly baptized in the Irish Episcopal Church, he stopped attending at the age of ten. In the Preface to *Immaturity* he recalls his childhood prayers; but it must be remembered that this is the recollection of a man of seventy-six, who has become one of the world's great story-tellers:

> In my childhood I exercised my literary genius by composing my own prayers. I cannot recall the words of the final form I adopted; but I remember that it was in three movements, like a sonata, and in the best Church of Ireland style. It ended with the Lord's Prayer; and I repeated it every night in bed. I had been warned by my nurse that warm prayers were no use, and that only by kneeling by my bedside in the cold could I hope for a hearing; but I criticized this admonition unfavourably on various grounds, the real one being my preference for warmth and comfort. I did not disparage my nurse's authority in these matters because she was a Roman Catholic; I even tolerated her practice of sprinkling me with holy water occasionally. But her asceticism did not fit the essentially artistic and luxurious character of my devotional exploits. Besides, the penalty did not apply to my prayer; for it was not a petition. I had too much sense to risk my faith by begging for things I knew very well I should not get; so I did not care whether my prayers were answered or not; they were a literary performance for the entertainment and propitiation of the Almighty; and though I should not have dreamt of daring to say that if He did not like them He could lump them (perhaps I was too confident of their quality to apprehend such a rebuff), I certainly behaved as if my comfort were an indispensable condition of the performance taking place at all.[8]

He goes on to say that he "continued these pious habits long after the conventional compulsion to attend church and Sunday School had ceased," but that eventually his "intellectual conscience" obliged him to refrain from such "superstitious practices."

Shaw's biographer, the late Archibald Henderson, once wrote me that if "Sonny" Shaw had not been removed from Church when he was ten, he would have become a faithful Church member instead of a recusant one. I cannot accept Pavlovian conditioning to such an extent. If Shaw had been kept in Church longer his separation might have been more explosive, but it would certainly have come. As it was, his baptism into the Church of Ireland was merely a mark of snobbish class-consciousness which destined him to a boyhood of genteel poverty with all the pretensions that went with it. Besides, I must object to any view of Shaw as a "recusant Christian" — a notion that keeps suggesting itself also throughout Anthony S. Abbott's generally useful book on *Shaw and Christianity*. Shaw, in his maturity, was not a recusant anything. He examined Jesus' teachings, as we shall see, and agreed with many of them, but in the end he had to write Jesus down as a failure; and he recommended not the reform of the Christian Church, but its abandonment in favor of a new religion that would "fit the facts."

In the 1870s the most popular and successful revivalists in the western world were the Americans, Dwight L. Moody (preacher) and Ira D. Sankey (song leader). In 1875, when Shaw was nineteen, the Moody-Sankey team invaded Ireland, and provided the first opportunity, so far as I know, for young G. B. Shaw to break into print. In a letter to *Public Opinion*, on the third of April, signed, simply, "S", the writer objected to an earlier correspondent's assessment of the evangelical revival. The revival was crowded, "S" maintained, because it was a free show, held in a building where admission was usually charged, and because Mr. Moody had "the gift of gab." The effect of the revival was to make individuals "highly objectionable members of society" and caused their "unconverted friends to desire a speedy reaction."[9]

The following year the young Shaw resigned his position with the real estate firm of C. Uniacke Townshend & Co., with whom he had been employed for the previous five years, and, like Larry Doyle, left Ireland for the larger world. His mother and two sisters had left two years before. The younger sister, in ill health, was placed in a sanitorium on the Isle of Wight, and died there a few days before George Bernard left Ireland. His mother and elder sister had established a household in London, leaving the son with a job that he did not like and a father who was an uncommunicative reformed alcoholic. It is not my purpose to turn biographical, but I must at least mention in passing the additional member of the Shaw household, who had preceded Mother and Sister to London, and whom they intended to join there — the charismatic music teacher, G. J. Vandeleur Lee. Lee had been part of a *ménage à trois* with the Shaws for some years in Dublin, and George Bernard was always at pains in his later years to assure his biographers that Lee's relationship with his mother was musical and platonic. Not all the biographers were convinced.[10] So far as can be determined, however, young Shaw's relations with Lee remained friendly, even to the point of an occasional collaboration with him in musical writings.

These relationships are of some interest here because, with the mass of literary remnants bequeathed to the British Museum at Shaw's death was a folder marked "Juvenilia," containing a forty-nine-page fragment of a "Passion Play" in blank verse (VII, 481–527).[11] It was written and abandoned in the second year after Shaw had come to London, when he was twenty-two. Its title is *The Household of Joseph* and it begins promisingly with a cast of characters headed: "Jesus, illegitimate son of Mary." Mary is a shrew, Joseph a harmless drunkard, and Jesus a renegade carpenter's apprentice who wastes his time lying in the fields spinning yarns for the spiritual edification of the townsfolk, until he is lured away from Nazareth to the big city of Jerusalem by a sophisticated stranger named Judas Iscariot, who brings him news, among other things, of John the Baptist. In Jerusalem Judas becomes something of a combination advance man and stage manager for the young preacher. He is unable, however, to keep his idealistic and rebellious prodigy under control, and when Jesus precipitates a street riot by driving the money-changers from the temple, Judas realizes that sooner or later Jesus will be in trouble with the authorities. He decides that it would be well for Jesus to meet a family he knows in Bethany — Martha, Mary, and Lazarus. At this point Shaw becomes confused (he was not the first) about the Biblical Marys. He makes Lazarus's sister Mary Magdalen, a well-known courtesan, part-time mistress to none other than Pontius Pilate. We are warned that Lazarus is an incurable drunkard, so we can anticipate what his being "raised from the dead" will entail. But we are spared that scene, as Shaw abandoned the script when he had gotten Jesus to Bethany.

The verse is more Shelleyan than Shakespearean. It is easy-flowing, but not especially musical or rich in imagery. There is no doubt that the youthful Shaw enjoyed making shockwaves by flaunting his independence from Christian mythology. But there is a symbiosis between the Judas and Jesus characters that might also have existed between Lee and Shaw, a teacher-learner, sophisticated-innocent relationship that makes them two sides of the same person. Jesus is certainly young Shaw, but Judas (very likely Lee) is also Shaw as he would like to picture himself at a more sophisticated stage. In a moonlit field near Nazareth, Judas advises the younger man:

> Then must thou
> Learn to stand absolutely by thyself,
> Leaning on nothing, satisfied that thou
> Can'st nothing know, responsible to nothing,
> Fearing no power and being within thyself
> A little independent universe.
> JESUS. But this is atheism.
> JUDAS. 'Tis so. What then?

Jesus then gives Judas the "first cause" argument: "Behold the world. Somebody must have made it." But Judas tells him, ". . . thou reasonest like a carpenter." Not everything is made. Some things, like the grain, simply grow. ". . . till this day," Jesus confesses, "I never met a man/ Who believed less than I. . . . But, were I to believe in no God at all/ I would, despairing die." (VII, 504–05) This heterodox argument is not resolved, but unquestionably Judas, for the moment, has the better of it.

The fact that Shaw refers to this fragment as a "Passion Play" indicates that he gave some thought to the unwritten scenes leading up to the crucifixion. He sensed that the climax of the drama ought to come in the confrontation between Jesus and Pilate — between the man of the spirit and the ruler of the temporal world. But in the scriptures when confronted by Pilate, Jesus refuses to defend himself, and the youthful Shaw, fresh from his years of watching touring companies play melodramas in the Dublin theatres, knew that a stage hero who refused to stand up to the villain and at least argue his point, would come off as arrogant, mad, or simply dull. Years later he advised an eminent Bible scholar, Henry B. Sharman (author of *Records of the Life of Jesus*), to abandon any attempt to put Jesus on the stage.[12] But when Shaw got hold of an idea he was like a dog with an old shoe and would not let go of it. At the age of seventy-seven, as a part of the Preface to his play, *On the Rocks,* he actually wrote the scene between Pilate and Jesus as it should have happened — that is, with Jesus vigorously talking back as Shaw would have done. It is about a fifteen minute scene (written this time in Shavian prose), and it ends with Pilate deciding

> You are a more dangerous fellow than I thought. For your blasphemy against the god of the high priests I care nothing: you may trample their religion into hell for all I care; but you have blasphemed against Caesar and against the Empire; and you mean it, and have the power to turn men's hearts against it as you have half turned mine. Therefore I must make an end to you whilst there is still some law left in the world. [VI, 625]

But in 1878 Shaw's Jesus was not so voluble nor so convincing. He had some of the same diffidence and some of the same arrogance given to Robert Smith, the autobiographical protagonist of Shaw's first novel, *Immaturity,* which he began writing about this same time. Indeed the most interesting aspect of *The Household of Joseph* may be the resemblance it bears, in the first scene, to the household of George Carr Shaw and Lucinda Elizabeth Shaw. The Shaw household, too, had been joined by a sophisticated outsider — J. G. Vandeleur Lee — and the young son was about to cross the Irish Sea to conquer his own Jerusalem or be crucified there. It is an adolescent image, but the images of adolescence can be, and often are, powerful and persistent.

The "first cause" argument between Jesus and Judas was a somewhat euphemized version of an actual discussion between Shaw and a Father William Edward Addis in a cell in the Brompton Oratory not long after Shaw first arrived in London. One of his new-found lady friends, concerned about the spiritual health of this self-proclaimed atheist, had arranged the meeting. It was Addis who said he should go mad if he lost his belief; but the unbeliever said he felt "quite comfortable."[13] At the urging of the worried young lady, Shaw agreed to wear a medal of the Virgin for a six months' trial. Neither the wearing of the medal nor Father Addis's persuasive arguments had any noticeable effect. Addis, by the way, was one of a number of sincerely religious men and women of the time who had difficulty in finding a spiritual home. He was a convert to Catholicism, but later left it to accept the pulpit of a nondenominational congregation in Australia, only to become, in his later years, an Anglican Professor of Old Testament at Oxford.[14]

Shaw, for the most part unemployed during his first nine years in London, and living off his mother's income from music lessons and a small annuity, nevertheless entered heartily into the life of the city. He found himself in the geographical center of a religious and intellectual turbulence that was spreading throughout the world. The city abounded in social and religious movements, and Shaw sampled many of them. How he was converted to economics by Henry George in 1882 and to the newly formed Fabian Society in 1884 is amply described by his biographers and by himself. Although from this time forward into the turn of the century, Shaw's overt devotion was to Fabian socialism, he knew that back of every movement as a driving force there had to be a religion. As he said later of Ruskin:

> He begins as a painter, a lover of music, a poet and a rhetorician, and presently becomes an economist and a sociologist, finally developing sociology and economics into a religion, as all economics and sociology that are worth anything do finally develop.[15]

And for himself, he had to confess

> I exhausted rationalism when I got to the end of my second novel at the age of twenty-four, and should have come to a dead stop if I had not proceeded to purely mystical assumptions.[16]

In 1888, writing under the name of Shendar Brwa, Shaw pretended to be a foreign visitor, ignorant of Christianity and completely confused by the lack of agreement among the British on the meaning of it.[17] In 1895 Shaw wrote to F. H. Evans: "I want to write a big book of devotion for modern people, bringing all the truths latent in the old religious dogmas into contact with real life — a gospel of Shawianity, in fact."[18] (Shaw had not yet taken the suggestion of his friend Sir Sydney Cockerell to substitute the *v* for *w* in derivatives of his name.) The following year his essay "On Going to

Church" appeared as the lead article in the first issue of *The Savoy.* In it he advised against going to *services,* where one's worship would be interrupted by priests and choir boys, but he accepted the location as favorable. Much of the essay is given over to the effects of church music and church architecture. "Any place where men dwell, village or city, is a reflection of the consciousness of every single man. In my consciousness there is a market, a garden, a dwelling, a workshop, a lover's walk — above all a cathedral"; and he asks of the builder to show him "where, within the cathedral, I may find my way to the cathedral within me." In modern civilization a church is a necessity:

> No nation, working at the strain we face, can live cleanly without public-houses in which to seek refreshment and recreation. To supply that vital want we have the drinking-shop with its narcotic, stimulant poisons, the conventicle with its brimstone-flavoured hot gospel, and the church. In the church alone can our need be truly met, nor even there save when we leave outside the door the materialisms that help us to think the unthinkable, completing the refuse-heap of "isms" and creeds with our vain lust for truth and happiness, and going in without thought or belief or prayer or any other vanity, so that the soul, freed from all that crushing lumber, may open all its avenues of life to the holy air of the true Catholic Church.[19]

2

Evolution of the Superman

The force that sweeps the characters of *You Never Can Tell* into the twentieth century is perhaps the most joyous force in all of Shaw, and it culminates in a basic repudiation of rationalism: The distinguished Queen's Counsel is asked for an opinion concerning the wisdom of the marriage of the penniless dentist, Valentine, to the wealthy "advanced" Gloria. "All matches are unwise," the Counsel intones. "It's unwise to be born; it's unwise to be married; it's unwise to live; and it's wise to die." "Then," interposes the Waiter, who happens also to be the Counsel's father, "if I may respectfully put a word in, sir, so much the worse for wisdom!" (I, 793).

Shaw (if we assume that the Waiter is speaking for the author) did not arrive at this position overnight. He had begun writing plays in the mid-1880s. By the end of the century he had written ten of them. He was by then well known in London as a music critic, drama critic, socialist pamphleteer, and platform personality; but he was not yet taken seriously as a playwright, despite the reasonably successful publication of the plays in three volumes. "It is clear that I have nothing to do with the theatre of today," he wrote to his beloved Ellen Terry at the end of the century: "I must educate a new generation with my pen from childhood up — audience, actors, and all, and leave them my plays to murder after I am cremated."[1] Fortunately he did not have to wait that long. In 1904 his playwriting career can be said to have begun in earnest with the Vedrenne-Barker management of the Royal Court Theatre in Sloane Square. Specifically, after the command performance of *John Bull's Other Island* for Edward VII, GBS became suddenly popular. During the years 1904 to 1907 eleven Shaw plays were performed at the Court for a total of 701 performances. All other authors combined totaled 287 performances. Once his popularity was established, the public characteristically discovered the older plays of the eighties and nineties, which were then successfully revived.

In the pre-twentieth-century plays there were already hints of Shaw's mature religious thought. Charles Carpenter has made them the subject of an extended study in his *Bernard Shaw and the Art of Destroying Ideals.* He finds the earliest ones "propaganda plays" and the moving spirit behind them all a need "to destroy accepted illusions."[2] In the three published volumes Shaw identified them as "Unpleasant Plays" (*Widowers' Houses, The Philanderer, Mrs Warren's Profession*); "Pleasant Plays" *Arms and the Man, Candida, The Man of Destiny, You Never Can Tell*); and "Plays for Puritans" (*The Devil's Disciple, Caesar and Cleopatra, Captain Brassbound's Conversion*). There is good sense in Carpenter's more specific classifications.[3]

I should like to divide them in yet another way: The first four are concerned with social and ethical questions — slum landlordism, individual morality, prostitution, militarism and romanticism — without heavy reliance on mystical qualities. The remaining six all imply that forces beyond rational explanation are at work in guiding human destinies. There is the secret that makes the poet "the stronger of the two" in *Candida.* There is Napoleon's "star" in *Man of Destiny.* In addition to the Waiter's quoted remark in *You Never Can Tell,* there is the young dentist who is hurled into matrimony almost against his will (foreshadowing Jack Tanner) with a woman whose feelings he cannot comprehend. There is Richard Dudgeon, the Devil's Disciple, surprising himself by his own sacrifice, declaring "My life for the world's future!" (II, 137). There is Caesar, leaving the young Queen behind as he departs for Rome: "It matters not; I shall finish my life's work on my way back; and then I shall have lived long enough ..." (II, 288). There is Lady Cicely's authority over Captain Brassbound, with a source beyond normal feminine charm.

But when Shaw was in his early forties, something happened — something of an inner spiritual nature. I do not wish to imply any specific "conversion" experience. There is nothing on the later side of this turning point that cannot be found in embryo on the earlier side. Still it is a perceptibly different Shaw who imposes himself on the new century, one who has gone beyond "the art of destroying ideals" to explore a personal cosmology — one might even say a theology.

Maurice Valency[4] suggests that at this time in the pursuit of his self-education, Shaw had been reading through the "B" volumes of the British Museum catalog — Blake, Bradley, Bunyan, Butler (Bergson was not yet available) — and there can be no doubt that he found support and inspiration in this part of the alphabet. F. H. Bradley[5] reflected the ideas of Hegel in more readable English. And Samuel Butler's *Luck or Cunning?*, first published in 1886, parallels almost precisely Shaw's later pronouncements on evolution in *Man and Superman* and *Back to Methuselah.*

But the appearance of *Man and Superman* in 1903 signaled more than a maturation of philosophical ideas hatched in the great Reading Room of the British Museum (his only alma mater, as he later said). E. Strauss finds him, at this time, writing on the edge of despair, "engaged in an attempt to save the remains of his crumbling hopes and beliefs by transferring them on to a new plane."[6] It is true that there are such notes of urgency in "Don Juan in Hell," the dream scene that constitutes the third act of *Man and Superman*, but the spirit of the play is buoyant and vigorous, not in the least despairing. Shaw's biographer, St. John Ervine, may have been nearer the mark when he attributed the new look to Shaw's "happy marriage" on 1 June 1898 to Charlotte Payne-Townshend. For the first time in his life, Ervine maintained, Shaw lived in a well-regulated house where decent meals were served on time.[7]

What should be especially noted is that the marriage came after an extended period of physical and psychological distress. Ever since his delayed initiation into sexual relations at the age of twenty-nine (when a voice student of his mother, a widow in her forties, finally seduced him), he had been carrying on a series of philanderings, some physical, some literary. For Shaw, the literary ones were more meaningful than the physical. The intimate correspondence with the actress, Ellen Terry, probably the most famous paper-romance in literary history, began in 1892. The correspondents saw each other over the footlights or from afar, and met on business terms as playwright-actress only after the literary passion had subsided. But the passion was real enough. A selective sampling should be convincing: They had seen each other across the auditorium at the opening of someone else's play.

25 March 1897

DEAREST AND EVEREST,

I could not go any nearer to you tonight (even if you had wanted me to — say that you did — oh, say, say, say, say that you did) because I could not have looked at you or spoken to you otherwise than as I felt; and you would not have liked that in such a host of imperfectly sanctified eyes and ears. I was on the point, once or twice, of getting up and asking them all to go out for a few moments whilst I touched your hand for the first time.

I *saw* the play — oh, yes, every stroke of it. There was no need to look at you. I felt your presence straining my heart all through.

Think of that, dear Ellen, even when you had that wicked, cruel, Indian-savageous, ugly, ridiculous plumage in your blessed hair to warn me that you have no heart. Ah, if only — well, nonsense! Good night, good night: I am a fool.

G. B. S.[8]

All that, Charlotte Payne-Townshend brought to an end — though the correspondence dwindled on until 1920, eight years before Ellen's death.

There is no doubt that Shaw was much pursued by women, and that he was highly susceptible. But his true heart was never in love-making. His remark to Ellen was not wholly facetious:

> It is certainly true: I am fond of women (or one in a thousand, say); but I am in earnest about quite other things. To most women one man and one lifetime make a world. I require whole populations and historical epochs to engage my interests seriously and make the writing machine (for that is what G. B. S. is) work at full speed and pressure: love is only a diversion and recreation to me.... I act the lover so diabolically well that even the women who are clever enough to understand that such a person as myself might exist, cant bring themselves to believe that I am that person. My *impulses* are so prettily played — oh, you know: you wretch, you've done it often enough yourself.[9]

He struggled always to be free of sexual entanglement to devote himself to the higher moral passion. The parallel of this struggle to that of John Tanner in *Man and Superman* has often been noted, most perceptibly by Charles Berst in his excellent *Bernard Shaw and the Art of the Drama.*[10] But Shaw did not struggle against the union with Charlotte. Rather he was much relieved. Nor was Charlotte anything like the sexually provocative Ann Whitefield of the play. Their marriage, as their friends seemed to know and accept, was a companionate one in which sex had no place. And although Shaw said, in his old age after Charlotte's death, that he probably never should have married, it is difficult to imagine that he could have produced either the quality or the prodigious quantity of creative work under any other sort of arrangement. One must judge the marriage, by any ordinary standards, as highly satisfactory.

Marriage meant freedom also — at least psychologically — from his mother, whom he had followed to London in 1876, after she had abandoned him and his father in Dublin four years earlier. From childhood on, the maternal bond was one of love-hate ambivalence.[11]

There were other disturbances. Less than two months before his marriage, Eleanor Marx, Karl Marx's talented and romantic daughter, committed suicide. Shaw knew her from the days when the British Museum Reading Room was a kind of club for penniless intellectuals; and he knew also her common-law husband, Edward Aveling, whose infidelities and irresponsibilities drove her finally to swallow prussic acid. Aveling himself died of a kidney ailment a few months later. Shaw was to recall the tragic relationship later in *The Doctor's Dilemma.*

Also he had been overworking and was not in good health. "Oh, Lord, Ellen," he wrote to the actress in August of 1897, "I am so tired, even in the morning. I get out of bed so tired that I am in despair until I have braced myself with tubbing. When I sit down my back gets tired; when I jump up, I get giddy and have to catch hold of something to save myself from falling."[12] The malaise contributed, perhaps, to a series of bicycle accidents — one of them serious enough to require stitches — and dental troubles necessitating the laceration of his gums. Through it all he continued his *Saturday Review* articles, his vestry meetings, and his Fabian appearances, in addition to his own writing and a growing amount of business arrangements concerning the publication and production of his plays — which were just beginning to bring him a significant income. Finally he was laid low by a mysterious infection of his foot in April of 1898. It was diagnosed as necrosis of the bone and required two operations to remove the infected portions. There was some talk of amputating one toe, but this proved unnecessary. Nevertheless he was on crutches for at least a year.

Charlotte, on being informed of GBS's condition, cut short her holiday in Rome, and returned to London at once to nurse him back to health. They were married in the Registry Office with a couple of close friends as witnesses. His new wife immediately swept him away from the vicissitudes of London to a cottage at Hindhead. It was the end of his career as a theatre critic, and a temporary respite from some of his other responsibilities. He had every intention of a "working honeymoon" but fate again stepped in. Unused to crutches, he misjudged the stairs at Hindhead, and in the tumble fractured his left arm. He recuperated at Hindhead, with an interval of summer sea-bathing at Cornwall, until September of 1899, when they were off, at Charlotte's insistence, on a six-week Mediterranean cruise.

Of course Shaw was not wholly idle during these long months. He wrote his book-length essay on Wagner's *Ring* (*The Perfect Wagnerite*), completed *Caesar and Cleopatra,* wrote *Captain Brassbound's Conversion,* and attempted to carry on his various enterprises by correspondence. Nevertheless the London pressures were definitely off, and in his wheelchair, and on the beach at Cornwall, and on the deck of the *Lusitania,* he had time and opportunity to think and reorder his life.

When the couple returned to London at the end of 1899 to settle in Charlotte's flat at Adelphi Terrace, Shaw was forty-three, Charlotte, forty-two. He had gained weight and was wearing good conservative clothes. Like the hero of the play he was about to write, his marriage had brought him into the establishment. Even after he had settled back into the London scene, the gestation period for *Man and Superman* was exceptionally long. In 1901 he wrote nothing but *The Admirable Bashville,* a blank-verse spoof of his early novel, *Cashel Byron's Profession* — hurriedly composed so that its copyright could forestall other dramatizations of the work. "Man & Superman no doubt sounds as if it came from the most exquisite atmosphere of art," he later wrote to his biographer, Archibald

Henderson. "As a matter of fact, the mornings I gave to it were followed by afternoons & evenings spent in the committee rooms of a London Borough Council, fighting questions of drainage, paving, lighting, rates, clerk's salaries &c &c &c; and that is exactly why it is so different from the books that are conceived at musical at-homes."[13]

Shaw later maintained that the first inkling of the play came to him as a scrap of dialogue: "I am a brigand: I live by robbing the rich. I am a gentleman: I live by robbing the poor,"[14] which found its place in the exchange between Tanner and Mendoza at the beginning of Act III. Despite these beginnings, it did not turn out to be another thesis play like *Mrs Warren's Profession*, but the culmination (as Alfred Turco, Jr., has pointed out) of everything since *Immaturity*, Shaw's appropriately titled first novel begun at age twenty. Turco has further pronounced it "the first play in which Shaw's belief in the possibility of an *effective* idealism is presented with real conviction."[15]

The play, however, was not quite the new Evolutionary Bible that Shaw, in retrospect, claimed it was: "I think it well to affirm plainly that the third act, however fantastic its legendary framework may be, is a careful attempt to write a new Book of Genesis for the Bible of the Evolutionists" (II, 532).[16] He later told a correspondent that "the 3rd Act of Man and Superman will remain on record as a statement of my creed."[17] And in a 1910 letter to Count Leo Tolstoy accompanying a copy of *The Shewing-Up of Blanco Posnet*, he says again: "It is all in Man and Superman, but expressed in another way."[18] All this might much better have been said of *Back to Methuselah*, which was still nearly twenty years in the future. As we shall see, *Man and Superman*, with all its crackling brilliance, is only a beginning of Shaw's exploration of the Life Force. In 1921 he recalled, more accurately, ". . . in 1901, I took the legend of Don Juan in its Mozartian form and made it a dramatic parable of Creative Evolution. But being then at the height of my invention and comedic talent, I decorated it too brilliantly and lavishly" (V, 338).

Man and Superman consists of what Shaw called a "trumpery story" — a sex-chase, really — surrounding an hour-and-a-half dream scene of Don Juan in Hell. As a practical man of the theatre Shaw realized that the complete play was not likely to achieve immediate production, even though it created considerable comment when it was published. Its first real stage run — minus the dream scene — replaced the successful *John Bull's Other Island* in 1905 at the Royal Court. It was immediately popular. The play, complete with dream scene, does make a unified whole, despite Shaw's prefatory apology that his play had "a totally extraneous act" and that you must "prepare yourself to face a trumpery story of modern London life, a life in which, as you know, the ordinary man's main business is to get means to keep up the position and habits of a gentleman, and the ordinary woman's business is to get married" (II, 503–04). Early critics took this offhand remark at face value, but Maurice Valency now finds that the dream scene is "a lens through which the love story is magnified to cosmic proportions,"[19] and Norman Holland correctly perceives that the Hell scene "casts long shadows into what preceded and what follows."[20]

As to the comedy itself, it is all very well to call it "a Victorian farce in four neatly arranged acts,"[21] or to suggest that without the Hell scene the play is a bright sexual comedy not unlike those of Terence Rattigan.[22] Even on these terms it outperforms all its models. In his merging of plot and counter-plot, in his use of triangles and reversals and the chase, in his elegant change of pace and character-contrast, and most of all in the subtleties that flow beneath the surface dialogue and inhabit the imagery — in all these ways Shaw, at the turn of the century, could give lessons to Eugene Scribe or any of the French farceurs, or to W. S. Gilbert or Oscar Wilde, or, for that matter, to the later Noel Coward or Terence Rattigan.

Samuel Butler, to whose thinking Shaw owed so much, made a happy figure of speech about the nature of evolution:

> As in the development of a fugue, where, when the subject and counter subject have been announced, there must henceforth be nothing new, and yet all must be new, so throughout organic life — which is a fugue developed to great length from a very simple subject — everything is linked on to and grows out of that which comes next to it in order — errors and omissions excepted.[23]

Man and Superman, which deals with evolution, is also such a fugue, and its subject is simply: *The Life Force, acting through the will of woman, subdues man to its purpose, and thereby moves the race to its next higher level.* That is the subject that gives unity to the play, even though it proves somewhat limiting to its philosophical development.

It is a subject that calls for a sex play. And Shaw, himself a Victorian as far as his public manners were concerned, a man who could not bring himself to utter or write a vulgar four-letter word, and writing for an audience reared in an era of sexual repression, still managed to dramatize an obviously sexual pursuit, climaxing in a mutual orgasm played front and center. For there is no doubt that in the "obligatory scene" in Act IV, where Ann has finally caught up with Tanner and sheer feminine attraction keeps him from fleeing the garden in Granada, the act of mating virtually ceases to be symbolic and becomes all but actual. It is here that the term "Life Force" is first used in the play proper. And A. M. Gibbs[24] calls our attention to the play on the word "will":

> TANNER ... Your father's will appointed me your guardian, not your suitor. I shall be faithful to my trust.
> ANN (*in low siren tones*) He asked me who I would have as my guardian before he made his will. I chose you!
> TANNER. The will is yours then! The trap was laid from the beginning.
> ANN (*concentrating all her magic*) From the beginning — from our childhood — for both of us — by the Life Force.
> TANNER. I will not marry you. I will not marry you.
> ANN. Oh, you will, you will. [II, 728]

Then there are echoes of the dream scene, and Ann, desperately throwing aside all her previous posing, says in anguish, "No. We are awake; and you have said no.... I made a mistake: you do not love me." Tanner, unprepared for this sudden release of tension, tumbles, as in jiujitsu, the other way: "(*seizing her in his arms*) It is false: I love you.... I have the whole world in my arms when I clasp you.... "

> ANN. Take care, Jack: if anyone comes while we are like this, you will have to marry me.
> TANNER. If we two stood now on the edge of a precipice, I would hold you tight and jump.
> ANN (*panting, failing more and more under the strain*) Jack: Let me go. I have dared so frightfully — It is lasting much longer than I thought. Let me go: I cant bear it.
> TANNER. Nor I. Let it kill us.
> ANN. Yes: I dont care. I am at the end of my forces. I dont care. I think I am going to faint.

And as the others discover them:

> ANN (*reeling, with a supreme effort*) I have promised to marry Jack. (*She swoons. ...*) [II, 729–30]

Does she really? Listen to Shaw instruct the first Ann Whitefield, Lillah McCarthy:

> Dont forget to say "But you nearly killed me, Jack, for all that" as if you meant it. He *has* nearly killed you. Mrs. Lyttleton, close behind me, explained to her party that Ann was only pretending to faint. That is not exactly true. Ann doesnt faint exactly, but she does collapse from utter exhaustion after her "daring so frightfully."[25]

That's as candid as Shaw would dare to be. One feels confident that Lillah understood.

How is it then, that a society that only a few years before rejected *Mrs*

Warren's Profession as scrofulous, accepted *Man and Superman* with such delight? There was in *Mrs Warren* no such provocative figure as Ann Whitefield, and no such physical encounter as the embrace of Jack and Ann. Louis Crompton has pointed out that emancipated Victorians — like Roebuck Ramsden in the play — had come to accept theological and political radicalism, but clung desperately to traditional sexual mores, despite such avant gardists as Havelock Ellis, Grant Allen, and H. G. Wells. Perhaps Shaw's most shocking heresy was to treat sentimental love, marriage, and child-bearing as though they were three unrelated phenomena, whereas in the Victorian-Edwardian code the three are, or at least ought to be, inseparable.[26]

The answer must be, at least in part, Margery M. Morgan's, that after all, the conventional idea of marriage wins out in the end, "leaving a philistine audience happily complacent,"[27] whereas the break between Kitty Warren and her daughter Vivie makes no such comforting compromise, and lays the blame for prostitution inevitably on a competitive capitalist economy. But there is, I think, another reason why *Man and Superman* could be accepted and enjoyed by the early Edwardians. Shaw had, in the meantime, mastered a style. Daniel J. Leary thinks that Shaw's characters, especially when they become most archetypal, behave like puppets.[28] And Margery Morgan finds them resembling the standard commedia dell'arte types. Both of these images would be useful to the Shavian actor. For although Shaw talks a good deal about "realism," he is a realist only in the philosophical sense that he wishes to face the real issues of life and of his times. In his view of theatrical behavior he is closer to Mozart and Italian opera and Charles Dickens and nineteenth-century melodrama than he is to Ibsen or Chekhov or the French Naturalists. Martin Meisel has settled this once and for all with his *Shaw and the Nineteenth-Century Theatre*. This does not mean a rejection of reality. It is a way of viewing reality from the safer distance of comedy. Shaw had inherited the knack of Aristophanes and Molière — and it is difficult to extend the list — of making audiences laugh and sending them home to think.

"This is a curious psychological thing," Shaw confessed when he was seventy-two years old:

> It has prevented me from becoming a really great author. I have unfortunately this desperate temptation that suddenly comes on me. Just when I am really rising to the height of my power that I may become really tragic and great, some absurd joke occurs to me, and the anticlimax is irresistible.... I cannot deny that I have got the tragedian and I have got the clown in me; and the clown trips me up in the most dreadful way. The English public have said for a long time that I am not serious, because you

> never know when the red-hot poker will suddenly make its appearance or I shall trip over something or other.[29]

In this way the harsh realities are made palatable and enjoyable, and even, perhaps, thought-provoking. But the realities are there none the less. Both women — Violet and Ann — bring their quarries to earth. Violet has succeeded with rapacious ease and is already secretly married and pregnant when the play opens. Her struggle, the counter-subject of the fugue, is merely to win the consent of her Irish-American husband's father and to secure an income from him. But the real excitement of the play is the sex-chase: the subtle entrapment devised by Ann — even, as it turns out, before the play begins — and carefully laid out in Ramsden's drawing room; the discovery in the park by Tanner that the trap has been set for *him*, and not, as he imagined, for his friend Octavius; the flight across the channel and through Southern Europe with Straker racing Violet's husband's American steam car (the love affair of Straker with the automobile being another and prescient variation of the sex-chase); the capture by brigands in the Sierra Nevada and the Don Juan dream among the outlaws; the overtaking by the English and American party and the final capitulation of Tanner. (What an excruciating sequence of experiences would all this have been for August Strindberg!)

To the sense of anti-climax of which Shaw found himself "guilty" must be added a sense of paradox, which is not quite the same thing. Shaw had a way of seeing things as a mirror-image — or perhaps since we are used to seeing ourselves reversed in the mirror, we should say a video-monitor-image — the shocking "reversal" of our own self-image, but as it is normally seen by others. A good paradox, in other words, is not merely a statement stood on its head regardless of the truth, but a surprising view from the other side. Such an inversion occurs when Don Juan Tenorio, the legendary pursuer of women, becomes John Tanner, the pursued.

Shaw claims that Tanner was modeled on H. M. Hyndman, the rousing Marxist leader of the Social Democratic Federation, and it is Hyndman who is described at Tanner's first entrance. The spirit of "The Revolutionist's Handbook" which Tanner is supposed to have written is not too different from Hyndman's 1883 manifesto, *The Coming Revolution in England:* "We come before you as Revolutionists, that is, as men and women who wish to see the basis of society changed."[30] Stanley Weintraub perceives Tanner as an analogue of Sidney Trefusis, the hero of Shaw's fifth novel, *An Unsocial Socialist.*[31] And undoubtedly there is a good deal of Shaw himself in the speech-making Socialist he is satirizing. In pretending to write his own biography as Frank Harris "ought to have done it," he confesses that "Tanner, with all his extravagances, is first hand: Shaw would probably not deny it and would not be believed if he did."[32] But in the end Tanner, like any good character from any good writer of fiction, is essentially himself, despite any family resemblances he may have to others.

The inversions are not limited to the Don Juan-Tanner character. When we find ourselves in Hell, we are not so much in a place as in an atmosphere, a state of mind. And this we must assume to be a significant aspect of Shaw's "creed." Shaw no longer considered himself an atheist — as he had announced himself in his youth — but he had no belief in any kind of personal immortality.[33] The "Hell" is recognizable only because it is introduced by Mozart's musical theme, because Don Juan assures the newcomer that that is where she is, and because it is supervised by a congenial Devil. These are Shaw's concessions to convention; but we are periodically reminded that both Heaven and Hell are constantly with us, that we continue to move from one state to the other, or that they are often mixed together. When Ana asks if she could go to Heaven if she wished, the Devil answers contemptuously, "Certainly, if your taste lies that way." And when she quotes Scripture: "Surely there is a great gulf fixed," the Devil explains,

> Dear lady: A parable must not be taken literally. The gulf is the difference between the angelic and the diabolic temperament. What more impassable gulf could you have? ... the gulf of dislike is impassable and eternal. And that is the only gulf that separates my friends here from those who are invidiously called the blest. [II, 647–48]

We experience these two states of being in our dreams and suppress them in our waking moments. Shaw would probably agree with Freud that we would all be healthier creatures if we were more frequently aware of these dream-states.[34] The scene, then, is specifically Hell, but we are given views of Heaven by the Statue of the Commander, who comes to visit the Devil and decides to stay; by Don Juan, who is uncomfortable in Hell and at the end chooses Heaven; and by the Devil himself, who admits that he is prejudiced.

In Christian theology God is Love. The inversion here is that Love is Hell. Hell is the home of all the maudlin romantic emotions — joy, happiness, beauty — the mention of which tends to make Don Juan ill. In Hell you abandon all hope, that is, you abandon all moral responsibility. You have nothing to do but amuse yourself. That is the secret of its popularity. To this list of Stygian attributes Richard M. Ohmann very perceptively adds *an absence of order.*[35] The passion for order is an underlying force in all of Shaw, reflecting his revulsion against the disorder of his own upbringing. It is easy to see the untidy and disorganized household of the later *Heartbreak House* as a kind of Hell. It is notable that none of these attributes have anything to do with punishment. Hell is the natural condition of useless, selfish, pleasure-loving people.

But what then is Heaven? It is "the home of the masters of reality," a place where "you live and work instead of playing and pretending" (II, 650–51). Most of all it is a place of contemplation. To the Devil and the the Statue all this is pretty dull — "the most angelically dull place in all creation." The Hell-dwellers view it much as the youth in the final play of *Back to Methuselah* will view the elders who are becoming "vortices of pure thought." "Oh, it suits some people," the Devil admits. "... it is a question of temperament. I dont admire the heavenly temperament.... but it takes all sorts to make a universe. There is no accounting for tastes" (II, 647).

The Hell scene, despite the burden of its verbiage, has proved, on purely theatrical terms, to be eminently playable. But its substance remains in the vigor of its argument. Shaw must be on the side of Don Juan and the Life Force, but here as elsewhere he gives the Devil his due. As the stage direction describes him he

> *is getting prematurely bald; and, in spite of an effusion of goodnature and friendliness, is peevish and sensitive when his advances are not reciprocated. He does not inspire much confidence in his powers of hard work or endurance, and is, on the whole, a disagreeably self-indulgent looking person; but he is clever and plausible, though perceptibly less well bred than the two other men, and enormously less vital than the woman* [II, 643]

(On whom is *he* modeled? Oscar Wilde, perhaps?)

At any rate his arguments are so forceful and so cogent that the result of the debate is left in doubt. As with Undershaft in *Major Barbara* and Cauchon in *Saint Joan,* the antagonist all but overpowers the cause which Shaw champions on the lecture platform. But the Devil is essentially a hedonist, and "Can there be any doubt," asks Crompton, "that his religion is the real religion of our cultivated middle classes and especially of those university teachers who are not mere philistines?"[36]

Turco also places the argument in a contemporary framework:

> Had Don Juan been depicted, say, as an intelligent Roman Catholic, and Lucifer as an existentialist, the central issue would be unchanged. That issue lies in the confrontation between the man who commits himself to working for what he believes to be life's larger purpose and the man who, seeing sceptically and perhaps rightly, concludes that there is no cause to think that existence has meaning and hence that any effort expended on its behalf is futile.[37]

What has all this to do with the emerging idea of a Life Force? The term itself is not mentioned until we are well into the dream scene, and its power does not become evident until the Ann-Tanner scene in Act IV. "It is curious," Berst observes, "how obliquely the Life Force applies to the play itself," and he is correct in relegating it to a "subliminal" motivation.[38] Even in the celebrated Preface there is more about the need for a genuine sex play and about the development of a literary style than there is about the Life Force. Similarly, though there are a few highly illuminating sentences in "The Revolutionist's Handbook," its basic theme is that improvement of the species can take place only through selective breeding. The brief Foreword to the popular edition of *Man and Superman* is somewhat more revealing, but this was not written until 1911.

Still, one must not underestimate the significance of the breakthrough in Shaw's thinking that *Man and Superman* represents. At the core of that breakthrough is the much-quoted announcement of Don Juan:

> I sing not arms and the hero, but the philosophic man: he who seeks in contemplation to discover the inner will of the world, in invention to discover the means of fulfilling that will, and in action to do that will by the so-discovered means. [II, 664]

Now what Shaw had so far discovered about "the inner will of the world" was that a mysterious Life Force was at work in directing evolution. He had not yet thought this through, and it would be another twenty years before he did so. For the present he could perceive that this Force favored organization as opposed to chaos, and that it required brains — intellect: "My brain is the organ by which Nature strives to understand itself." The Life Force says to the philosopher:

> "I have done a thousand wonderful things unconsciously by merely willing to live and following the line of least resistance: now I want to know myself and my destination, and choose my path; so I have made a special brain — a philosopher's brain — to grasp this knowledge for me as the husbandman's hand grasps the plough for me. And this," says the Life Force to the philosopher "must thou strive to do for me until thou diest, when I will make another brain and another philosopher to carry on the work." [II, 684–85]

The Life Force, however, has no brain of its own. Dependent as it is, for the moment, on human brains, its immediate task is to evolve a being superior to the present inadequate specimen. Since this can only be accomplished through the process of mating, the sex play is central in its exalted purpose.

It is to woman that Nature has given the responsibility for propagating the race, and she is therefore, whether she knows it or not, more in the service of the Life Force than man is.

> Sexually, Man is Woman's contrivance for fulfilling Nature's behest in the most economical way. She knows by instinct that far back in the evolutionary process she invented him, differentiated him, created him in order to produce something better than the single-sexed process can produce. [II, 659]

But the male sex, once created, fulfilled its breeding purpose and had considerable energy left over. All of civilization is a sublimation of this energy, which has resulted in woman herself being dominated by man, and being forced to cooperate in "a feeble romantic convention that the initiative in sex business must always come from the man" (II, 506).

This notion, traceable to Bradley or Lester Ward, is something of a stumbling block to those who like to think of Shaw as a champion of women's equality, and seems to be in contradiction to his profession that "I have always assumed that a woman is a person exactly like myself."[39] In *Village Wooing* the Man tells the Woman: "I am a woman; and you are a man, with a slight difference that doesnt matter except on special occasions" (VI, 541). But the "slight difference" turns out to be an all-pervasive one in Shaw's evolutionary hypothesis.

There are cases — cases of genius — in which the creative flame burns as brightly in the man as in the woman. One gathers that the hope for the future lies with those who *know* they are in the service of the Life Force. "The true joy in life [is] being used for a purpose recognized by yourself as a mighty one" (II, 523).

This was about as far as Shaw had gone by 1903. There has been much scholarly speculation as to how he came by these ideas. (There has also been considerable challenge from the scientific community, but this will be dealt with in Part II.) Certainly Butler's *Luck or Cunning?* was behind them, but Butler's "cunning" is not exactly attributed to the Life Force. Valency sees a good deal of Schopenhauer in them, though he thinks that Shaw may have got his Schopenhauer second-hand. For Schopenhauer, behind the world of observable scientific phenomena there existed an urge which was neither rational nor intelligible. He called it "the will to live" — a phrase perhaps closer to "the Life Force" than Bergson's later *élan vital*. One might almost have thought that Schopenhauer, writing in 1844, had the future Ann Whitefield in mind when he observed that ". . . the sexual impulse, although in itself a subjective need, knows how to assume very skilfully the mask of an objective admiration, and thus deceive our consciousness; for nature requires this stratagem to obtain its ends."[40] Shaw, however, did not share Schopenhauer's longing for Nirvana, or the sense that life was both painful and meaningless. In his general attitude he would have sympathized more with Schopenhauer's rival, Hegel. Carl Henry Mills thinks that Shaw must have been influenced by Lester Ward's *gynaecocentric* theory which first appeared in the 1880s. Ward's theory, as its name implies, rests on the original dominance of the female, man's eventual subjugation of her, and man's building the world with his excess time and energy.[41]

There were quite evidently many sources of such ideas available to Shaw by 1900. Shaw himself acknowledged — with notable reservations — his debt to Nietzsche's *Übermensch*. We know he was impressed with Carlyle, and that he had just completed his study of Wagner's *Ring*. At various times he took pains to acknowledge also his debt to John Bunyan, Voltaire, Shelley, Dickens, Ibsen, the writers of the Gospels, and (at a later date) Henri Bergson. John Gassner recalls the Max Beerbohm cartoon in which the jaunty young nineteenth-century Shaw offers some old clothes for sale to the noted European critic Georg Brandes standing behind a counter as "a merchant of ideas."

> "What'll you take for the lot?" asks Brandes.
> "Immortality," says Shaw.
> Brandes protests: "Come, I've handled these goods before! Coat, Mr. Schopenhauer's; waistcoat, Mr. Ibsen's; Mr. Nietzsche's trousers—"
> To which Shaw blandly replies, "Ah, but look at the patches."[42]

Shaw's idea of a Life Force was not original. But neither was it a copy. The patches supplied by Shaw make the doctrine of creative evolution and the search for social and economic justice into a unified garment. It is the synthesis that is original. Shaw had an unusual facility for absorbing all kinds of diverse material, refining and amalgamating it into something that could legitimately be called his own. It could hardly, at this point, be called a religious statement. He came closest to such a statement in "The Revolutionist's Handbook" when he pictured the political man as a failure, looking to the heavens for help and finding them empty. "He will presently see that his discarded formula that Man is the Temple of the Holy Ghost happens to be precisely true, and that it is only through his own brain and hand that this Holy Ghost, formally the most nebulous person in the Trinity, and now become its sole survivor as it has always been its real Unity, can help him in any way" (II, 742).

I must differ here with Charles Berst's view that "there is a stasis in religion incompatible with the dynamic art of Shaw's exploratory dialectic."[43] Though a Shavian religion has not yet fully emerged in *Man and Superman*, certainly we are beyond the time when we must assume that "there is a stasis in religion." It is precisely Shaw's point — to be made again and again in the following decade — that religion *must* be dynamic to survive — a point that apparently still needs to be made today.

Ann Whitefield would not have understood any of this, nor would she have bothered to try. She is not an advanced or liberated woman. She is, as Margery M. Morgan has labeled her, a virtuoso player of the game within the rules of Victorian society.[44] To her the "Life Force" sounds like the "Life Guards." "I doubt if we ever know why we do things," she tells poor Octavius. "The only really simple thing is to go straight for what you want and grab it" (II, 716). Since the only solid job the Life Force has to do in this play is to bring about the mating of Ann and Jack, we may well ask, "Why all the fuss?" Why is not the mating of Violet and young Malone an equal matter of Life Force concern? And why does the Life Force not take the much easier way to a healthy offspring and mate Ann with the willing Octavius rather than with the struggling Tanner? The dramaturgical answers are simple, of course: the easier solutions would not make an exciting play. The philosophical answers are less obvious.

It is a legitimate criticism of the play certainly that after all the philosophical build-up of the dream scene about discovering "the inner will of the world," the only result that that will can produce is a fairly conventional marriage. It will have deeper and more far-reaching implications for Father Keegan, Barbara Undershaft, Blanco Posnet, Androcles' friend Lavinia, and Joan of Arc. But at this point Count Leo Tolstoy is justified in concluding that "This book of yours expresses your views not in their full and clear development, but only in their embryonic state."[45] Here, then, in its infancy, we must say that the Life Force's primary interest is simply in breeding; that it favors the high vitality and sexual drive of Ann over the haughty, business-like Violet, and the intellectual vigor of the apostle of change over the sentimental idealist. The union of Ann and Jack combines, as Berst points out, instinct with intellect — "an epiphany in the higher consciousness of intuition."[46]

But it is through Ann that the Life Force flows. I am not sure I would go so far as to agree with Morgan that the play is *about* the primacy of the female will,[47] but certainly without Ann's drive there would be no play at all — even though Tanner knows, perhaps subconsciously, that the childhood love-pact between them remains inviolate. Ann is the *efficient* organ of the Life Force.[48]

It is important to point out that no "superman" appears in this play. Perhaps — but only perhaps — the Life Force is at work trying to create one, or at least move a step closer to one. At the end of the dream scene when Ana asks the Devil, "Tell me: where can I find the Superman?" his answer is unequivocal: "He is not yet created, Señora." (And Arthur Nethercot reminds us that it is only near the end of the dream scene that the word "Superman" occurs in the play.)[49] As the scene fades, Ana, who has chosen neither Heaven nor Hell, decides that her work is not yet done, and cries out to the universe for "A father for the Superman!" (II, 689).

In a later program note (1907) Shaw explained that "Though by her death she is done with the bearing of men to mortal fathers, she may yet, as Woman Immortal, bear the Superman to the Eternal Father" (II, 803).[50] When Octavius repeats to Ann a sentimental speech we have already heard in the dream scene, Ann has "a strange sudden sense of an echo from a former existence" (II, 678, 714). May we not also conjecture that Ana's final cry has a similar resonance in Ann's unconscious?

From the beginning, the critical reactions to *Man and Superman*, despite its popularity, were mixed. Max Beerbohn could not find a play in the printed volume (though he was later enthusiastic over *John Bull's Other Island*).[51] A. B. Walkley, to whom the play was dedicated and the long preface addressed, found the "idea-plot" not substantive enough to nourish the "action-plot."[52] William Archer wrote, "We see very clearly in *Man and Superman* one of the main reasons why Mr. Shaw will never be an artist in drama. It is that his intellect entirely predominates over, not only his emotions, but his perceptions."[53] The actress, Mrs. Pat Campbell, once remarked that Shaw was all "fireworks and ashes," and some later critics, such as Allan Rodway, continue to agree with her.[54] He finds Shaw's comedy "dehumanizing." Berst, too, finds "the beauty of the rhetorical pyrotechnics more satisfying than the logic of the discourse."[55] But he acknowledges that "the sum of its parts is ... greater than the parts themselves," and that "the total play is remarkably complex and effective, incorporating more quintessential Shaw than any other."[56]

If there are those who find less than is there, there are others who possibly find more. Valency declares it to be "the most ambitious [exposition of a dramatic idea] anyone had so far put forth in the theatre." He also discovers in the play a "symbolist" drama after the manner of LugnéPoë since it affords "an intimation of the spiritual reality which lies, presumably, beyond the sensual experience."[57] Ervine judges it "as a piece of craftsmanship possibly the most remarkable comedy that has ever been written."[58]

When Albert Einstein replied to Bernard Shaw's toast at a London banquet in 1930, he was speaking of Shaw's plays in general, but he might well have been thinking of *Man and Superman* specifically:

> Whoever has glanced into this little world sees the world of our reality in a new light. He sees your puppets blending into real people so that the latter suddenly look quite different from before. By thus holding the mirror before us, you have been able as no other contemporary to effect in us a liberation and to take from us something of the heaviness of life.[59]

The idea of a Life Force, thus launched, continued to absorb more and more of Shaw's thoughts until, after the end of World War I, he returned to deal directly with it in the most extensive and probably the most misunderstood of all his works, *Back to Methuselah.* In the meantime there are no more plays specifically about the Life Force, but we can continue to sense its presence and discern some of its attributes by observing the characters who are most likely to be in its service.

The Life Force, it turns out, is no respecter of creeds. We have seen how, in *John Bull's Other Island,* it produces a radical variant of orthodox Catholicism in the defrocked priest who finds the composure of a dying Hindu on a spiritual level beyond the reach of his Christian ministrations. But it is just as likely to infect a Major in the Salvation Army. Unlike John Tanner, who theorized about poverty and social inequality, Barbara Undershaft faced these problems daily in the Salvation Army shelter. She thinks of herself as a born-again Christian until she confronts another and disturbing aspect of the Life Force — her estranged father, a munitions magnate who has overcome poverty by manufacturing weapons of destruction. Thus in *Major Barbara* Shaw challenges religion with the social problems of both poverty and war.

Under the weight of Undershaft's challenge the fundamentalism of the Salvation Army (whose vitality Shaw genuinely admired) proves as inadequate for Barbara as orthodox Catholicism had for Peter Keegan. She is forced to agree with her father that a religion that accepts poverty and is content with palliative measures is not helping Life in its upward struggle. Barbara's great disillusion comes when she perceives that the great spiritual Army to which she has given her life is really at the mercy of distillers and armament manufacturers — at the mercy, that is, of an amoral (and often immoral) capitalist system. What shocks Barbara into new hope is the discovery that her father, "the Prince of Darkness," is not the great pragmatist he is taken for, but a mystic like herself. Cusins, the unlikely Greek scholar who is to marry Barbara and take over the arms factory, aptly observes to Undershaft, "You do not drive this place: it drives you." And when he asks, ". . . what drives this place?" Undershaft gives the surprising reply, "A will of which I am a part" (III, 169).

Undershaft's profession is morally questionable, but he has built an environment of order, cleanliness, and economic sufficiency. His workers do not need handouts from the Salvation Army. Are they further from the Kingdom of God or closer to it? Barbara is puzzled, but no longer dismayed. At this point she joins her creator, Shaw, in the search for a new religion:

> I have got rid of the bribe of bread. I have got rid of the bribe of heaven. Let God's work be done for its own sake: the work he had to create us to do because it cannot be done except by living men and women. When I die, let him be in my debt, not I in his; and let me forgive him as becomes a woman of my rank. [III, 184]

Major Barbara remains one of Shaw's most controversial plays. We know from his correspondence with Gilbert Murray (the Adolphus Cusins of the play) that it also caused him much anguish in the writing.[60] He had allowed his characters to create a moral dilemma which could not be satisfactorily solved within the context of the play. Shaw rejected the temptation of melodramatic manipulation in favor of a more honest ending — one that satisfied the immediate aspirations of the main characters but left the moral question unresolved. He could find no easy virtue in an immoral society.

Other attributes of the Life Force emerge: It is not always rational, nor does it necessarily support conventional morality. It is apparently on the side of Jennifer Dubedat (*The Doctor's Dilemma*) in valuing the life of a talented artist who is, by all conventional standards, a scoundrel over the life of a useful but unimaginative medical doctor of impeccable social behavior. It opposes the ritualized marriage of Edith and Cecil in *Getting Married*, and disapproves of the entire set of strictures that surround the institutions of legal mating.

It operates with enigmatic power. It transforms the rascally horse thief, Blanco Posnet, into a Good Samaritan before he knows what has happened to him. It is definitely on the side of Lena Szczepanowska in *Misalliance* — that is, on the side of risk-taking and adventure and a straight-forward, honest, day-to-day approach to life. And it is on the side of faith and the future — the simple faith of Androcles, but more especially the sophisticated faith of Lavinia — rather than the brute strength of Ferrovius, the rationalism of the Captain, or the temporal power of the Emperor.

It is, in other words, against stasis and for change. It accepts Alfred Doolittle's degeneration in *Pygmalion* from a fascinating dustman into middle-class anonymity, just as it glories in the transformation of Eliza from a grubby flower-girl into a charming and useful "lady."

William Archer to the contrary notwithstanding, Shaw was a consummate dramatic artist. In creating his plays he took obvious enjoyment in role-playing his own antagonists. The Devil in *Man and Superman* is completely un-Shavian in his hedonism, sentimentalism, and cynicism, but he is consistently entertaining and forceful. Andrew Undershaft, unlike the Devil, is hard-working, pragmatic, and forward-looking. Yet there is something Satanic about him too — a power for Death, which, if unleashed, could negate all the vitalist and spiritual qualities toward which Shaw sees the Life Force striving.

But Shaw's own dramatic genius kept getting in the way of the Life-Force message, just as it had earlier got in the way of his Socialist message. The creative dramatist continues to be fascinated by the Undershafts, the

Devils, the Napoleons, the Cauchons — those who stand in the way of evolutionary progress toward the Superman, as well as those who carry its banner. And indeed, within the context of the dramas, it is not always obvious which is which.

If Shaw was to develop a true religion of the Life Force, it was clear that he could not wholly entrust it to the recalcitrant characters that grew out of his imagination. He would have to put his artistry aside, from time to time, and mount the platform in person.

3

Preacher of a New Religion

The most vocal and powerful rebels against organized religion in post-Darwinian England were the members of the National Secular Society. Militant atheists, they were banded together under the leadership of Charles Bradlaugh, who, after five years of stormy litigation, became the first avowed atheist to sit in the House of Commons. There has been nothing quite like the National Secular Society before or since. Unlike most of the reform groups of the seventies and eighties, it was not a haven for dissatisfied intellectuals. The membership was largely working class, and its meetings resembled those of the early labor unions more than those of a debating society. Their regular meeting room was in the Hall of Science, and science (though most of them knew little about it) was the new god. These were men and women who had shaken off the shackles of their former superstitious faiths and looked forward to the dawning of a new age. Most of them were not socialists. Bradlaugh himself was, in fact, vigorously anti-socialist, believing that secularism produced an independent mind that would not yield to any kind of collectivism. There were other freethought societies in London in those decades, but none of the others had comparable leadership. Bradlaugh, up from the London streets, self-educated in the law, handsome and magnetic, was really a kind of atheistic revivalist preacher. He was joined, in the mid-seventies, by England's greatest woman orator, Annie Wood Besant, and together they carried secularism to new heights.

But Bradlaugh pushed his own exertions too far; his health broke, and in January of 1891, at the age of fifty-seven, he died. Annie Besant had, in the meantime, deserted Secularism, first for socialism, and then for Theosophy. The NSS was left leaderless. Much as a bereaved congregation tries out new ministers before selecting a permanent one for its pulpit, the Secularists began inviting likely candidates to address them at the Hall of

Science. In the months following Bradlaugh's death, they got round to Shaw. Shaw had supported Bradlaugh in his effort to be seated in Parliament, but he had otherwise little in common with the Bible-smashing tactics and science-worship of the Secularists. But he accepted the invitation to speak, and had, as he later recalled, "an exceedingly pleasant evening." Nearly twenty years later he recalled:

> I do not think it would have been possible for Bradlaugh to have thrown the most bigoted audience of Plymouth Brethren into such transports of rage as I did the freethinkers at the Hall of Science. I dealt with the whole mass of superstition which they called free thought: I went into their Darwinism and Haeckelism,[1] and physical science, and the rest of it, and showed them that it did not account even for bare consciousness. I warned them that if any of them fell into the hands of a moderately intelligent Jesuit — not that I ever met one — he could turn them inside out.[2]

Shaw then proceeded to give them his interpretation of the Trinity ("You are the father of your son and the son of your father") and the Immaculate Conception ("I believe in the Immaculate Conception of Jesus's mother, and I believe in the Immaculate Conception of your mother").

The speech to the Secularists must have been all the more disconcerting because at the time it was delivered Shaw was still the brilliant young Irish iconoclast who could be counted on to upset the establishment. Instead Shaw treated *them* as an establishment — as fundamentalists of the secularist movement. His remarks, at least as he remembered them, were saucy and provocative, but he offered little in the way of a positive philosophy beyond the idea that mechanistic atheism was not enough.

The year after he wrote *Major Barbara*, however, he began making speeches that might properly be called heterodox sermons. He continued such preaching as long as he was in public life, and, via radio, into his old age. When he appeared in the pulpit of the City Temple (*not* at a Sunday service) it was a Shaw the audience had not anticipated:

> A little uncommon without being startling [a reporter noted], rather tall, fair, and bearded, he was attired in a dark suit, a white turn-down collar, and a meek sort of tie. He uses pince-nez which for most of the evening dangled in front of his double-breasted, close-buttoned jacket. His voice is a little metallic, but not unpleasing, his enunciation is very distinct, with barely a suspicion of an Irish accent.[3]

Some two weeks later, when Shaw spoke to the Christian Socialist Guild of Saint Matthew on "Some Necessary Repairs to Religion" the reporter for the *Clarion* expanded the description to include the audience.

> You can get a tooth drawn for five shillings, with gas, and judging from the torture suffered by a number of people present, I should think they would have preferred to have a tooth drawn, without gas, than to have their cherished religious convictions torn out by the author of *You Never Can Tell.*
>
> There was no anesthetic property in Mr. Shaw's "gas." He used a well-worn pair of forceps, and they hurt. I was sitting at right angles to most of the patients, and I could see their faces. They were an interesting study, so fascinating that at times they captured my whole attention and I missed one or two of the lecturer's brilliant coruscations.
>
> One man literally squirmed in his seat. His facial muscles were twisted, his teeth ground together, his hands nervously gripped his coat. And yet he seemed to feel that the dentist was quite right. The tooth was a bad one. It must come out.
>
> One dear old lady nodded her head in dissent for an hour. An old man sat and glared with an expression of gloom which gradually deepened into despair. A jovial young man tried to assume an air of smiling, good-natured contempt, but the smile went all awry at times. His tooth had begun to ache. I felt sorry for the people, although the experience will do them good.[4]

The same sort of reports continued. Seven years later, back at the City Temple, the *Christian Commonwealth* again noted the response:

> More than once the audience sat still and breathless, fascinated by his quite terrifying earnestness and by the merciless vigor of his attack upon the dearest delusions and pretences by which we buttress our self-esteem. The light, patronizing laughter with which half-educated people receive Mr. Shaw's attempt to shock them out of their deadly moral complacency was not heard at the City Temple. The atmosphere was inimical to laughter. And the sight of his tall, tense figure in the pulpit, electrical in its suggestion of vital energy completely under the control of his will — even his beard of flax (to use Richard Middleton's phrase) had a peculiar quality of liveness — compelled a similar intensity of interest and attention from his hearers. Laxity, either of mind or body, is impossible when Mr. Shaw is speaking. Several times I looked round upon my fellow-auditors to mark the effect of his words. I saw consternated faces, hostile faces, faces which bore an expression of alarm and even horror, but not one suggesting boredom or mere tolerance of an inexplicable perversity on the part of the speaker.[5]

According to Blanche Patch, his secretary in his later years, Shaw never wrote out his speeches, although he prepared meticulously for them and often made copious notes which he did not take to the lecture platform. He realized that a spoken style should not be the same as a written one. A performer always, he wanted to play upon his audience, utilizing their laughter, judging for himself the duration and intensity of their attention, varying his tempo and attack accordingly. In these lectures he never spoke as a mere entertainer, but always with the tone of high purpose.

It is unlikely, however, that he followed strictly the extempore method described by Blanche Patch in those instances when he prepared a formal paper for the Fabian Society. In 1906 he delivered such a paper on the subject of Darwin and Darwinism as part of the Fabian series on "Prophets of the Nineteenth Century." The "notes" which he preserved from this lecture constituted a complete essay on the subject, obviously intended for publication, but somehow never published. The double-spaced typed manuscript that survives was placed in a folder along with the notes from his earlier essay, *The Quintessence of Ibsenism*, and the pages were renumbered by hand. The Darwin paper begins at the renumbered page 46 and ends on page 99. Six assorted pages are missing. We can only surmise how much of the original 54-page essay was read from the lecture platform.[6]

Fifteen years later he drew heavily on this previous work in writing his Preface to *Back to Methuselah*. The missing pages in the manuscript were probably absorbed bodily into the new Preface. Emendations on the typed sheets appear in Shaw's longhand (and at a few places in Pitman shorthand). One must assume that they were made when he was reviewing the material in 1921. The typed portion of these notes, then, can be taken as an expression of Shaw's thought on evolution in 1906, and can serve as a basis for comparison with his more mature views.

Certainly the basic pattern of the *Methuselah* Preface is already apparent in the 1906 notes. By this time Shaw has adopted both Lamarck and Samuel Butler. Indeed, a sizable portion of the lecture is given over to a review of Butler's life and works, with some emphasis on the difficulties posed by Butler's intransigent personality. Butler was the first man of genius to realize the moral abyss that Darwin had opened up, but because he could not help turning his horror into a personal attack on Darwin, and because of his often perverse sense of humor, Butler found that his ideas were largely ignored or condescended to by the scientific establishment. It is clear that Shaw considered himself a more amenable, and, he hoped, a more persuasive surrogate for Butler.

There is no evidence here that Shaw had studied Lamarck as carefully as he had studied Butler. It was enough that both of them regarded the *Will* as a source of creative energy. Such energy has little place in the evolutionary pattern of Darwin, and none at all in that of Darwin's followers, whom Shaw labeled the Neo-Darwinists. Many of these Neo-Darwinists found it simpler to worship the originator of Darwinism than to think the idea through for themselves. As Shaw put it,

> We play about on a dust heap of gunpowder called the world without the slightest suspicion that it is anything but inert dirt. Then some studious old gentleman comes along and throws down a lighted match and bang! we find ourselves whirling in a nebula of flame and tempest. Under such circumstances it is extremely hard not to look with considerable awe on the person who threw the match, and even to conclude that it was he who miraculously changed the earth from dirt into dynamite.[7]

What emerges from these notes more clearly than from any other of Shaw's writings is his need for a religion to complement his Socialism. Both Darwinism and Marxism, however, fail to meet his need because they rest on presumptions of inevitability, which leave them fatalistic and life-denying:

> Marx's value theory, his fatalism, his materialism, his atheism, his calm omission of human character, passion, will, or even consciousness as factors in social evolution, his conception of class war as a discord which will finally resolve itself into the economic concord of Socialism: all these are as early Victorian as they can be; and this is the reason why Marx, when he had once destroyed the moral prestige of Capitalism, was of no further use to the Socialists, and forces [sic] them to get rid of Marxism as the first condition of their advance, exactly as Darwin, when he had once destroyed our belief in the book of Genesis, is forcing us all to get rid of him also as the first condition of regaining our spiritual energy.[8]

Neither Darwinians nor Marxists, therefore, can really be good revolutionists, striving to reform society, so long as they accept this position of inevitability. Rather, Natural Selection under Capitalism is surely more likely to favor those whose fundamental rule in "Thou shalt starve ere I starve."[9] Shaw had recently given this same line to Andrew Undershaft in *Major Barbara* (III, 173) and it presents a good example of how Shaw's political and religious ideas found their way into his plays. We can no longer find it surprising that although in his lecture he cites such a motto as being in opposition to the Life Force, he places it, as a dramatist, in the mouth of a highly complex human character who makes his own claim to a mystical source of power. Life, as Shaw dramatized it, was never as unambiguous as it appeared in his lectures and expository writings.

Shaw does warn his audience, however, not to accept his message too simplistically. He is neither a rationalist nor an "irrationalist." "What is wrong with [the Neo-Darwinian]," he tells them, "is not that he believes in

Natural Selection, but that he believes in nothing else."[10] Certainly we should make all possible use of our reason. But for the past half century Science has abandoned metaphysics and has relied on reason alone. This leaves out the humanizing influence of mythology and the truth that is inherent in all myth. Consequently Science has tended to attract unthinking people who cannot, for instance, conceive of the possibility that "all habits are both acquired and inherited"[11] — the crux of the argument between the Darwinians and the Lamarckians. The nineteenth-century mind "had been so bent towards logical and mechanical considerations by false religious teaching, that at last [it] became incapable of mystical truth."[12]

This applies to that body of myth called the Bible. Simply because the Bible proved to be unacceptable as a scientific tract, we should not therefore completely disregard it.

> The Bible as an epic of the will of God struggling tragically to fulfill itself through the agency of its grotesquely imperfect vessel Man, and sometimes driven in its agony to remonstrate with him through the mouth of Balaam's ass, is, with all its childishness and tribal idolatries, enormously more real and impressive than any body of rationalist literature that has yet been produced.[13]

In his peroration Shaw told the Fabians essentially what he told his audiences at the City Temple and at the Guild of St. Matthew:

> . . . we must make a religion of Socialism. We must fall back dogmatically on our will to Socialism, and resort to our reason only to find out the ways and means. And this we can do only if we conceive the will as a creative energy, as Lamarck did, and totally renounce and abjure Darwinism, Marxism and all fatalistic, penny-in-the-slot theories of evolution whatsoever. . . . We have to conceive this will which inspires us with the purpose of establishing Socialism as being what everybody used to call the will of God, and what some of us now prefer to call the purpose of Life. For our personal ends this will or purpose cares very little. . . . It makes you act as if you believed it could work miracles; and this is an extremely uncomfortable and even ridiculous course of conduct unless you actually do believe in the miracle of Lamarckian evolution. To attain and maintain such faith, you must attach yourself, not to the reasonableness of the miracle, but to the unquestionable fact that it actually does occur. It has occurred in all our lives. Take my own case, and see whether it is not essentially yours also. My life has been a miraculous transformation of a good-for-nothing boy into the writer of this paper and of several quite unaccountable uncommercial plays, through a mysterious will in me which has prevailed over environment, heredity, and every sort of external discouragement. What is more, that will is not me: it makes the merest instrument of me — often

> overworks and abuses me most unreasonably. It makes me perform the feats of a bold, energetic, resourceful man, though I am actually a timid, lazy, unready one. It makes me write things before I understand them; and I am conscious that my own subsequent attempts to explain them are sometimes lame and doubtful. There can be no doubt that all writers who are original or inspired — whichever you choose to call it — write down things which are seen by later generations to imply a good deal that the writer himself would have vehemently denied. It is this will that has made you and me Socialists; and what it has done to us it can do to England and to the world.[14]

Written in Shaw's hand after this passage is the note, "End here if too late to go on," but the notes continue for five more pages.

The Life Force, acting as *Will*, prodding evolution along through its various stages, and emerging in our time as a shaper of human destiny — this was the germinal idea behind *Man and Superman*. But whereas there it was content to mate a couple of vitalists in the hope of producing an offspring that would be a step closer to the superman, here, in this speech, as in other speeches and sermons before World War I, the Life Force kept enlarging its scope. As we shall observe in Part II, the process was to culminate in the creation of Shaw's own myth stretching from the Garden of Eden to "as far as thought can reach."

In this pre-war decade, however, he had only just discovered Henri Bergson, and he had still to look more deeply into Schopenhauer, as he came to connect the evolutionary process ever more closely with the *Will*. He would, along the way, cease using the term "Natural Selection" in favor of the more accurate "Circumstantial Selection" — a change which has gone unnoticed by the scientific community. He was to give further thought both to the chronology of evolution and to the possibility of speeding its progress. He had not yet considered the possibility of voluntary longevity. He had only thought briefly about the ultimate question of what happens if the Life Force fails in its human experiment — if "Man is not up to the mark." At length Shaw was to find the concept of creative evolution universal enough to embrace politics, education, science, and art — to be a repository of hope, even after the near-destruction of European civilization in 1914–1918. For the present it was still inhibited by its too-close ties to Socialism, but it at least lighted a path out of the wilderness of materialism and determinism.

The lecture circuit offered Shaw the opportunity to be more informal and more frankly exploratory than he could be in his plays; but in 1909 one of his sermons spilled over almost directly into a play. Perhaps because of this it remains one of the strangest works to come from his pen. On the surface *The Shewing-Up of Blanco Posnet* is a short takeoff on the American Western. It has had few rivivals. It is too short to play for a full evening and too long to be considered a normal one-act. Shaw says it is a religious tract in dramatic form, but the Lord Chamberlain thought the religious ideas shocking and refused it a license. Shaw thereupon sent it to the Abbey Theatre in Dublin, where it was performed on a bill with other short plays. London did not see it until after World War I, when censorship became more relaxed. A contemporary viewer would be hard put to discover what it was the Lord Chamberlain objected to.

The opening directions call for dress and speech "of pioneers of civilization in a territory of the United States of America," but no Western American ever said, "Bully for you," or "Not likely," or "I shan't blab" (to mean "I won't squeal"). Certainly not one of Shaw's major works, it is nonetheless a lively and amusing play, ending with the disreputable horse thief, heretofore proud of his wickedness, humbly confessing that he went soft and risked his own life by giving the stolen horse to save an unknown woman's child. He has no explanation for this conduct except that God "plays cat and mouse with you. He lets you run loose until you think you're shut of him; and when you least expect it he's got you." Under the circumstances Blanco Posnet, acquitted, feels called upon to preach a sermon, and it is the sermon for which the play is the excuse. Incomprehensibly, it is the sermon that was the principal reason for the Lord Chamberlain's censorship.

At his own unsophisticated level, Blanco has to deal with the problem of evil in the world. After he had stolen the sheriff's horse, and then given it away to a woman who had to get her child to a doctor, the child died of croup; so his one good deed had gone for naught.

> What about the croup? It was the early days when [God] made the croup, I guess. It was the best he could think of then; but when it turned out wrong on His hands He made you and me to fight the croup for Him. You bet he didnt make us for nothing; and He wouldnt have made us at all if He could have done His work without us. ... He made me because he had a job for me. He let me run loose till the job was ready; and then I had to come along and do it, hanging or no hanging. [III, 797]

Although the Life Force, unlike the conventional concept of God, is not omnipotent, it occasionally invades the conscience of such an unlikely agent as Blanco Posnet — or Dick Dudgeon — and compels him to do its will. The "inspired" character becomes more and more familiar in Shaw's later plays. But such characters were not always aware that they were being used for Nature's higher purpose, and, once created and set free in Shaw's imagination, they could not always be trusted to deliver a clear Shaw message. When one of them is forced to preach an obvious Shaw sermon — as Blanco is forced to do — the artistry of the play suffers.

One of Shaw's devices for preserving his own artistic integrity is the preface. No previous playwright had ever given so much consideration to the published play — that is to say, to the general reading public. No doubt Shaw's concern for the play-reader was motivated by the fact that his early plays did not immediately find their way to the stage, or, if they did, it was to the stage of a small experimental theatre. From the very beginning Shaw dressed his plays up for the printed page, making sure that the reader could visualize the people, the setting, and the action, and removing all evidences of stage jargon, such as "Exit rear right center," or "Cross R. upstage of sofa." Consequently his plays attained a readership far more extensive than his live audiences. And since the plays did not always carry his complete message, the extended prefatory essay became a standard part of the publication.

The voice of the sermonizer can frequently be heard in these prefaces. In the Preface to *Man and Superman* it is cajoling, iconoclastic, brilliant. In the Preface to *Androcles and the Lion* (one of his longest), it is more sombre, but the tone of the preacher is even more evident. The war clouds were darkening Europe, and they cast their shadow over the play that begins as a somewhat adult children's pantomime and ends with a religious message. In his later years Shaw was to experiment with the "serious farce," but *Androcles and the Lion* is simply a farce that gets lost in seriousness. With its vaudevillian prologue and its hands-and-knees lion it has understandable trouble in delivering its more profound message. Nevertheless, its Christian martyrs are unforgettable in their variety, from the meek Androcles through the powerful Ferrovius and the conscience-ridden Spintho to the noble Lavinia. And if audiences can stop laughing at the final antics of the Lion, the Emperor, and Androcles in the Roman arena, they can hear Ferrovius making his choice to join the Pretorian Guard: "The Christian god is not yet. He will come when Mars and I are dust." And Lavinia making hers: "I'll strive for the coming of the God who is not yet" (IV, 634).

The play itself had, as we shall note, a mixed reception. As Shaw was preparing it for publication Europe went to war. The debacle of 1914–1918 was to present him with a spiritual crisis that would profoundly challenge his notions of human progress and put his faith in the Life Force to a severe test. It is at first thought hard to conceive of the Preface to *Androcles* as being written just after — almost simultaneously with — *Common Sense About the War*, the pamphlet which brought all patriotic Britain down on his head, and made him, until tempers cooled, the most hated man in England. A second consideration, however, will reveal that the two essays are in many ways complementary. Both have to do with what Shaw called the "Barabbasque" attributes of our civilization.

The Preface comes as close as he ever came to writing the "Gospel of

Shawianity" that he had proposed in 1895.[15] Since he had been brought up in a non-religious household, Bible-reading had not been forced on him, but the Bible was one of the books available to him, and he read through it at an early age — as far as the Pauline Epistles, at which point "I broke down in disgust at what seemed to me their inveterate crudeness of mind."[16] He became fonder of the Bible as he grew older and came to know parts of it quite well. In his late years he and Charlotte read from it to each other frequently.

In the course of writing *Androcles* he had taken himself back to the early Christian era, and that imaginary journey led him to take a fresh look at the New Testament, particularly the Gospels. His commentary on the Gospels, which forms the largest part of the *Androcles* Preface, is that of an intelligent mind responding directly to the printed page. There is no pretense of having looked into "sources" or scholarly exegeses. What he finds believable he looks at from his own twentieth-century point of view. What he cannot believe, he simply passes by without quibble.

Shaw notes at once that Jesus's countrymen chose Barabbas — a man of sedition and violence, though we know nothing else about him — over Christ. Through the centuries we have continued to choose him, though we have given him Jesus's name and ironically call his followers Christians. "I am no more a Christian than Pilate was, or you, gentle reader," Shaw begins disarmingly:

> and yet, like Pilate, I greatly prefer Jesus to Annas and Caiaphas;[17] and I am ready to admit that after contemplating the world and human nature for nearly sixty years, I see no way out of the world's misery but the way that would have been found by Christ's will if he had undertaken the work of a modern practical statesman. [IV, 459]

We cannot really call Christ's way a failure, since, except for the little groups like the Plymouth Brothers, no society has ever tried it. In this respect the Crucifixion was an unqualified political success for the Roman establishment: it eliminated Christianity. Looking over the history of the West from the days of Androcles to Europe in flames in 1915, Shaw delivers one of his most damning (and beautifully turned) indictments:

> . . . we pass our lives among people who, whatever creeds they may repeat, and in whatever temples they may avouch their respectability and wear their Sunday clothes, have robust consciences, and hunger and thirst, not for righteousness, but for rich feeding and comfort and social position and attractive mates and ease and pleasure and respect and consideration: in short, for love and money. [IV, 467]

In the Gospels Shaw finds much that would be familiar to an adherent of any non-Christian sect: the divine parentage, the ritual of eating (or drinking) the god, the insistence on faith, and the post-mortem rewards and punishments to keep the followers in line. All the evangelists, Shaw believes, belong to the first post-crucifixion generation, since they all seem to imply that the Second Coming will take place within the lifetimes of those then living. In which case, Shaw adds wryly, they must also "have believed that reading books would be one of the pleasures of the kingdom of heaven on earth" (IV, 492). (Shaw does not take into account the likelihood that the Gospels had been transmitted orally for a number of generations before they achieved written form.)

His distinction among the four Gospel writers is shrewd. Matthew draws a class-conscious aristocratic Jesus. He tells us the entire basic story, including some material that is irrelevant. Mark, Shaw finds, "adds nothing" except for the Ascension. Bible scholars would more likely say that it was Mark who told the basic story, and that the others added a great deal.

Luke is the artist, who tells the story for the sake of telling it. Shaw warms to Luke as a kindred artistic spirit, but he is put off by Luke's obvious enjoyment of melodrama and sentimentality. "His logic is weak; for some of the sayings of Jesus are pieced together wrongly, as anyone who has read them in the right order and context in Matthew will discover at once" (IV, 500). Shaw notes also that there is a strong feminine interest in Luke that is not present in the other Gospels.

John, too, has great literary charm. He is the only chronicler who claims to have been a witness to the events he is chronicling. Shaw finds him "educated, subtle, and obsessed with artificial intellectual mystifications," but like Matthew he treats the entire career of Jesus as having "no other purpose than to fulfill the old prophecies." The character of John's Jesus is "hopelessly irreconcilable" with those of the other evangelists (IV, 503).

Perhaps there is nothing in these readings of the Gospels that would impress a biblical scholar, but Shaw does pose some interesting challenges to orthodox Bible-reading, and I would recommend the *Androcles* Preface to any fundamentalist television sermonizer, and ask him at least to address the questions Shaw raises. Shaw finds in the Gospels the seed of a credible modern religion. But he rejects the Christ who was barbarously slain as a sacrifice and atonement, and the whole notion that we are saved from hell fire by our faith in this myth. He prefers instead the story of a prophet who, despite the fact that he came to believe himself to be "a crude legendary form of god" (IV, 561), gave us guides for behavior and political organization which we desperately need, and which we have steadfastly refused to try. For it is only by political means, says Shaw, that Christ's doctrines can be put into practice.

Shaw sees the essential doctrines of the Gospels as these: 1) The kingdom of heaven is within you, within all of us. We are therefore members one of another. 2) Get rid of private property. Shaw regards Jesus's injunctions against material goods as an endorsement for the communal ownership of all wealth. 3) Get rid of judges, punishment, and revenge. 4) Get rid of family entanglements.

Now these ends can be brought about only by an entire revision of our social and political systems. Central to all four is the second, the economic tenet, and Shaw devotes more than ten percent of his preface to the various options for dividing the world's resources equitably among the world's peoples — beginning with the population of Great Britain. (These same options, considerably expanded, he later repeated in *The Intelligent Woman's Guide to Socialism and Capitalism* in 1928.)

From his earliest readings of the New Testament, Shaw had trouble with Saint Paul. Twenty-seven years before he wrote the *Androcles* Preface, when "Jack the Ripper" was terrorizing London's East End, the editorial pages of *The Star* were lamenting the failure of Christianity which these murders represented. Shaw submitted a letter signed "J.C." asking

> Why do you put the Whitechapel murders on me?... As to the eighteen centuries of what you call Christianity, I have nothing to do with it. It was invented by an aristocrat of the Roman set, a university man whose epistles are the silliest middle class stuff on record. When I see my name mixed up with it in your excellent paper, I feel as if nails were going into me — and I know what that sensation is like better than you do.[18]

(The letter was not accepted by the editor!)

It is not surprising that Paul, in the *Androcles* Preface, continues to be the villain. Shaw recognizes him as a man of genius, but "under the tyranny of two delirious terrors: the terror of sin and the terror of death, which may be called also the terror of sex and the terror of life" (IV, 546). Paul did not in the least understand Jesus. His "conversion" was merely a fascination for Jesus's personality and a discovery of a way by which his twin terrors could be grafted on to Jesus's message.

> The great danger of conversion in all ages has been that when the religion of the high mind is offered to the lower mind, the lower mind, feeling its fascination without understanding it, and being incapable of rising to it, drags it down to its level by degrading it. [IV, 547]

Paul was a magnificent organizer and apparently a persuasive speaker and correspondent. By using Jesus's crucifixion as expiation for the terrors of sin and death, Paul built a very popular religion.

> ... to this day Pauline Christianity is, and owes its enormous vogue to being, a premium on sin. Its consequences have had to be held in check by the worldlywise majority through a violently anti-Christian system of criminal law and stern morality. [IV, 553]

So for twenty centuries the West has been following a Pauline Barabbas under the comfortable pretense that it has been following Jesus.

> Barabbas is triumphant everywhere; and the final use he makes of his triumphs is to lead us all to suicide with heroic gestures and resounding lies. Now those who, like myself, see the Barabbasque social organization as a failure, and are convinced that the Life Force (or whatever you choose to call it) cannot be finally beaten by any failure, and will even supersede humanity by evolving a higher species if we cannot master the problems raised by the multiplication of our own numbers, have always known that Jesus had a real message, and have felt the fascination of his character and doctrine. [IV, 515–16]

Shaw, like many another, took from the Bible what he wanted to take from it. "Belief," as he says, "is literally a matter of taste" (IV, 510). It is natural that Shaw should have found in the Gospels support for Fabian Socialism. He also found a spirit of Jesus that was compatible with the Life Force. But it is obvious, after reading the Preface to *Androcles,* that he could never think of himself as a member of any established sect. Nor, despite his fascination with certain aspects of Jesus's character and teachings, could any church accept him into its communion.[19]

The "new religion" that Shaw was preaching was specifically not any form of Christianity as it was espoused or practiced. But by 1915 it had attained some recognizable tenets. Let me try, in a few paragraphs, to summarize them as Shaw was expounding them at the beginning of the First World War:

Religion is a necessity. A divine purpose must be recognized in the universe. Life as a haphazard accident in the evolutionary process is too horrible to contemplate.

All present-day institutional religions are unsatisfactory. Though much of Jesus's personal theology is sound, no modern state would permit it to be put into practice. Institutional Christianity is a failure because the Christian Church is still caught between the Old Testament horror gods, and the intimidating salvationism of St. Paul.

True religion is always mystical. It is carried on by prophets, not priests. The true "protestant" renounces all churches and all priesthoods. The true "catholic" recognizes the communal relationship of all life.

Religion must be practical. It must concern itself with justice and economics and the social order and the divine value of life, as opposed to the otherworldliness of the medieval church. But sheer pragmatism as a rule of life cannot be tolerated. A practice is not right simply because it can be made to work.

God is not an omnipotent personality, but a blind Life Force, struggling through evolution and whatever other means are available to it to develop what one day might be the Godhead.

The Life Force has a will of its own, and its will is in men and women, perhaps in all living things. Geniuses are conscious of this purpose and try to help it along. Ordinary people may be doing its will unconsciously, but the true joy in life is in the knowledge of being used for purposes beyond the selfish desire for happiness.

The Life Force needs human beings to carry out its purpose. It needs our hands and brains. If we do not do well enough it will scrap us and develop a more satisfactory species.[20]

4

Interlude with a Bishop and a Dancer: "Mr. Shaw on Morals"

Androcles and the Lion opened in London on the first of September 1913. It produced cries from the press of "sacrilege" and "decadence" — though there were some favorable reviews, too. It closed after eight weeks.

Shaw was never reticent about talking back. The letters columns of the London papers are strewn with his acerb comments on every conceivable issue. He was particularly communicative on the subject of stage morals and censorship. The Lord Chamberlain was, until 1968, required to license every public production in London. In addition to *The Shewing-Up of Blanco Posnet*, Shaw had been refused permission for the presentation of two other plays, *Mrs Warren's Profession* in 1898, and *Press Cuttings*, a one-act play about Suffragettes, in 1909.[1] And although *Androcles* had not been banned, it was attacked on moral and religious grounds. Shaw was understandably sensitive on this point.

To anyone familiar with *The Philanderer, Mrs Warren's Profession, Man and Superman*, or *Getting Married*, Shaw's views on sex and marriage would come as no surprise. But the theatre-going public (or the preface-reading public) is not the same as the newspaper-reading public. And now, quite fortuitously, a wider opportunity for a discussion on morality presented itself — an opportunity to argue his points with one of London's leading Bishops in London's most prestigious newspaper, without reference to his own work. For one week *The Times* of London was forced to devote a daily column to letters under the heading: MR. SHAW ON MORALS.

It happened like this: On the same evening that *Androcles* opened, a popular French variety artiste, Gaby Deslys, opened a show called *A la Carte* at the Palace Theatre, a variety house under the management of

Alfred Butt. Gaby's show was a standard revue, apparently in no way remarkable except for her own glamorous presence in it. Even so, it had trouble running, and after a couple of weeks some changes were made. "The Gaby Glide has been added," *The Times* reported, "but even the strenuousness of that remarkable tangle could not conquer the irrepressible Mlle. Deslys or her equally energetic dancing partner, Mr. Harry Pilcer, last night."[2] There were the usual comedians and chorus.

It will be entertaining, perhaps, if not very instructive, to know something of the catalyst. Gaby Deslys was used to creating sensations. She was a slim, well-modeled blonde from Marseilles who made no pretense to great artistry. Her real name was Madeline Caire but on stage she had become "Gabrielle of the Lilies." She loved audiences and they loved her, to the extent that she left a fortune of several million francs when she died of a throat infection in 1920 at the age of thirty-six. She was noted, among other things, for her stage wardrobe and the way she wore it — and, apparently, on occasion, for the way she did not wear it — and she became in her time familiar with police action and censorship in a number of localities including, not surprisingly, Boston.

Her lure, however, must have been something more than that of a common leg-artist. Prior to the overthrow of the Portugese government, the youthful King Manuel, so gossip said, showered jewels upon her with scandalous disregard for the state coffers. And at her death she was mourned by the Duc de Crussol, who had followed her to Paris at her final illness and who fled afterwards to America to forget his sorrow. These reports are in addition to such casual evidence as the necessity for the police, on one of her American tours, to remove forcibly from the stage two Yale students in pursuit of her.

Now apparently the Lord Chamberlain's representative, in compliance with British law, had routinely witnessed the original show and the changes. But about the time when the Gaby Glide was added to the show, some complaints were made to local clergymen that Gaby's revue was indecent. The clergy ostensibly avoided going to the theatre themselves, but they later claimed to have made "careful and independent" inquiries and came to the conclusion that the performance was "suggestive" (which was to become a key word in the argument). Eventually the Church of England had to do its duty and expose to Gaby and her troupe an actual clergyman, Dr. W.S. MacGowan of St. Anne's, Soho, who attended, pen in hand, and — literally — made a list. The list was sent with a letter of complaint to the Lord Chamberlain over the signature of the Bishop of Kensington himself, Dr. John Primatt Maud.

With what seems from this distance remarkable deference to the clergy, the Lord Chamberlain's office dispatched the MacGowan list to Manager Butt with instructions that all the items on it be "rectified," and with a warning that his show would henceforth be kept under constant surveillance. Then, from the Lord Chamberlain's office, Col. Sir Douglas Dawson wrote the clergy a note of personal reassurance.

Here, fortunately for us, the clergy made a serious mistake in judgment. Presuming victory a little prematurely, they published Sir Douglas's letter, though it is clear he had not intended it for the public. Mr. Butt, therefore, very reasonably insisted that his reply to the Lord Chamberlain should also be made public, and the issue was then open for comment.

Manager Butt turned out to be embarrassingly articulate. He took up a number of the MacGowan points: "The incident of Mlle. Deslys powdering her legs (which, by the way, are fully covered) is purely a piece of light-spirited tomfoolery, and so far as I know has never been objected to by anyone." And the Gaby Glide, a descent of the stairs, though not in the Manager's own opinion graceful, certainly was not indecent.

To which, a few days later, Gaby herself added a touch of sentimental melodrama. "I know the admirable work that clergymen do. I know how they go into the slums and do their duty in the squalor of wretched hovels in the midst of revolting scenes and surroundings." Well, why, she wanted to know, didn't they come and see *her*? Why didn't they see her show or at least come and talk to her? She assured *The Times* readers that respectable people came to the Palace; they neither held up their hands in horror nor walked out on the show. "My dancing is acrobatic dancing; the same twists and turns occur in it as acrobats use in their tricks; certainly the same as done by hundreds of other dancers. If I do them with more brio than most, it is because I am enjoying the fun of it as well as the audience."

The Bishop of Kensington prepared his reply to the Palace Theatre with care and it was a full week before it appeared in print. In the meantime he took the precaution of sending the Rev. Dr. MacGowan back to the Palace, this time with a layman "whose long experience and tried judgment in such matters entitles him to the greatest respect"! Their mission was to determine whether in fact the performance *had* been changed and the public consequently "protected." The report of these gentlemen was that although the structure and dialogue remained unaltered, the "objectionable" parts had been altered or eliminated. Though this might sound equivocal to some, it was apparently by this time enough for the Bishop, who assured the public that since Monday the twentieth of October the changes requested by the clergy had been made.

The Bishop took the opportunity to say something about the motives behind all this. Certainly he and his fellow clergy had no personal prejudice against Mlle. Deslys. For more than a year they had been feeling uneasy about the tendency of the variety houses to introduce "undesirable and often highly dangerous" incidents into sketches after they had been sanctioned by the Lord Chamberlain. They used the Palace incident to call public attention to the "suggestive representations which made their appeal to the sensual and passionate instincts...." They were particularly concerned about the Young, whom they pictured as wandering into such theatres all unsuspecting.

At last, on the eighth of November, appeared a long letter from Bernard Shaw beginning, "May I, as a working playwright, ask the Bishop of Kensington to state his fundamental position clearly?" The Bishop responded promptly, and the debate was suddenly elevated into first-class intellectual fireworks. Letters poured in from all sorts of respondents. Most of the eighteen that were published dealt not at all with what went on at the Palace, but with what Mr. Shaw had said. The abrupt discontinuance of the column at the end of the week was obviously imposed by the editors.

In the temporal world of London the Bishop won his victory. The enticing Gaby was banished. Her show had closed and Mr. Butt reluctantly issued a statement that he had released Mlle. Deslys so that she could sail for America. ("Released?" remarked the dear old lady in *Punch* the following week. "I had no idea the impudent little baggage had been sent to prison. I hope it will be a lesson to her.")

As one who believed himself to be in the service of the Life Force, Shaw had to insist on an *individual* morality. It could not be the morality of the masses, for the Life Force is the spirit of change and the search for improvement. In the battle of moralities, therefore, it was Shaw against the field. Spurred on by the recent puritanical attacks on *Androcles,* he was both stimulated and prepared. He boldly set the terms and the rest of London supported or refuted.

I have tried to organize the debate under four general queries.

1. *Is not some form of censorship inescapable?*

Almost all the correspondents, including those of the theatrical profession, assumed that it was. H. B. Irving, writing from the Savoy Theatre, could only conclude: "There must be some ultimate authority... efficient in its organization, methodical in its inspection, and drastic in its regulations." Henry Arthur Jones, noting that it was impossible to censor "gestures, looks, by-play" in advance, actually advocated the establishment of an office of Inspector General, which would place observers in every theatre for every performance. Even the Bishop doubted the efficacy of that.

Elsewhere Shaw, too, had accepted the impartial enforcement of decency laws (as opposed to the caprices of the Lord Chamberlain's censorship), but in the present exchange he left the control of theatre morals wholly in the hands of the public:

> I venture to suggest that when the Bishop heard that there was an objectionable (to him) entertainment at the Palace Theatre, the simple and

> natural course for him was not to have gone there. That is how sensible people act. And the result is that if a manager offers a widely objectionable entertainment to the public he very soon finds out his mistake and withdraws. It is my own custom as a playwright to make my plays "suggestive" of religious emotion. This makes them extremely objectionable to irreligious people. But they have the remedy in their own hands. They stay away. The Bishop will be glad to hear there are not many of them. . . .

But the Bishop did not want to risk the public's contamination. He insisted that the censor had a job to do. And it was his "fundamental position" simply that sketches should not be altered after the Lord Chamberlain had approved them. On this single point he would have been happy to limit the argument.

> I do not wish to obscure that position by yielding to Mr. Shaw's invitation to follow him in a discussion of those paradoxes which to any normally thinking man carry their refutation on the surface.

Shaw found the Bishop's preoccupation with regulations instead of morals simply "an astonishing episcopal reply."

2. *Is anyone qualified to judge what is "suggestive," "objectionable," "undesirable," "highly dangerous"? Does not such presumption infringe upon individual rights?*

Shaw accused the Bishop of using words like "suggestive" and "objectionable" as if there were a general agreement as to their meanings.

> On the face of it the Bishop of Kensington is demanding that the plays he happens to like shall be tolerated and those which he happens not to like shall be banned. He is assuming that what he approves of is right, and what he disapproves of, wrong. Now, I have not seen the particular play which he so much dislikes; but suppose that I go to see it tonight, and write a letter to you tomorrow to say that I approve of it, what will the Bishop have to say? He will have either to admit that his epithet of objectionable means simply disliked by the Bishop of Kensington, or he will have to declare boldly that he and I stand in relation of God and the Devil. And, however his courtesy and modesty may recoil from this extremity, when it is stated in plain English, I think he has got there without noticing it. At all events, he is clearly proceeding on the assumption that his conscience is more enlightened than that of the people who go to the Palace Theatre and enjoy what they see there. If the Bishop may shut up the Palace Theatre on this assumption, then the Nonconformist patrons of the Palace Theatre (and it has many of them) may shut up the Church of England by turning the assumption inside out. The sword of persecution always has two edges.

. . . .

> By "suggestive" the Bishop means suggestive of sexual emotion. . . . The suggestion, gratification, and education of the sexual emotion is one of the main uses and glories of the theatre. It shares that function with all the fine arts.
>
> . . . .
>
> . . . the Bishop and his friends are not alone in proposing their own tastes and convictions as the measure of what is permissible in the theatre. But if such individual and sectarian standards were tolerated we should have no plays at all, for there never yet was a play written that did not offend somebody's taste.

But the Bishop denied any intention to impose his personal tastes on the public. He believed that "suggestive" and "objectionable" *could* have generally accepted meanings. And obviously he did not mean suggestive of the kind of sexual emotion engendered by contemplating works of fine art!

> I think further that I can assert that had Mr. Shaw himself seen the incidents in the performance at the Palace Theatre to which we directed protest, and which were promptly eliminated and altered, even he would have found difficulty in discovering there the suggestion of that kind of "sexual emotion" which he describes as "one of the main uses and glories of the theatre."

("As to that, I can only say," Shaw concluded in his final letter, "that if the Bishop sets out to suppress all the institutions of which I disapprove, he will soon have not one single supporter, not even/ Yours truly. . . . ")

In spite of the quip one must feel that neither Shaw nor the Bishop was on very solid ground here. It is true that whether Shaw would have approved of the performance or not was wholly beside the point. On the other hand, the Bishop might properly have asked if Shaw's analogies between sex and the arts would have been so exalted if the Puritan had been capable of looking at a really lewd performer.

The argument had spread from the press to the pulpit, and when the Bishop of London devoted his sermon to supporting the "vigorous campaign for a clean, pure life," Shaw characteristically applauded that sentiment with enthusiasm. However . . .

> [The Bishop of London] is reported to have declared that, "It has been said that no Christian Church has any right to criticize any play in London." It may be that there exists some abysmal fool who said this. If so, he was hardly worth the Bishop of London's notice. The Christian Church ought to criticize every play in London; and it is on that right and duty of criticism, not on the unfortunate Lord Chamberlain, that the Christian Church ought to rely, and, indeed, would rely without my prompting if it were really a Christian Church.

And this leads us to the more penetrating questions.

3. *Can the theatre ever be wholly separated from its religious impulses? Does the Church continue subconsciously to regard the theatre as a rival?*

It was to questions of this nature that the Bishop did not wish to "yield," assuming them to be mere "paradoxes." Few readers today, I think, would consider them unrelated to censorship and stage morals, but in 1913 only one correspondent (a Ronald Campbell Macfie) challenged Shaw's pronouncements linking the theatre with the church and he very briefly. For one thing, the matching of sexual intemperance to religious intemperance was too outrageous to bear comment. For another, very likely none of the debaters wished to risk further public hassle with the Great Contender, then fifty-seven and at the height of his powers both in print and lecture hall, on a subject in which he was considered heretical but expert — his City Temple lectures as well as a growing number of scenes and prefaces having already established him as a brilliant critical thinker about religion. "The theatre," he said, "is the Church's most formidable rival in forming the minds and guiding the souls of the people."

So closely related are the Church and the stage that for every objectionable feature, for every area of bad taste in the one, there is a counterpart in the other.

> I must remind the Bishop that if the taste for voluptuous entertainment is sometimes morbid, the taste for religious edification is open to precisely the same objection. If I had a neurotic daughter I would much rather risk taking her to the Palace Theatre than to a revival meeting. Nobody has yet counted the homes and characters wrecked by intemperance in religious emotion. When we begin to keep such statistics the chapel may find its attitude of moral superiority to the theatre, and even to the publichouse hard to maintain, and may learn a little needed charity. We all need to be reminded of the need for temperance and toleration in religious emotion and in political emotion, as well as in sexual emotion.

(It was here that Macfie took exception. Although he did not defend "religious extravagance," he felt we owe it to religious liberty to tolerate it. Not so with immorality. "Religious tolerance is one thing, moral tolerance is another.") Shaw continues:

> But the Bishop must not conclude that I want to close up all the places of worship: on the contrary, I preach in them. I do not even clamor for the suppression of political party meetings, though nothing more foolish and demoralizing exists in England today. I live and let live. As long as I am not compelled to attend revival meetings, or party meetings, or theatres at which the sexual emotions are ignored or reviled, I am prepared to tolerate them on reciprocal terms; for though I am unable to conceive any good coming to any human being as a set-off to their hysteria, their rancorous

bigotry, and their dullness and falsehood, I know that those who like them are equally unable to conceive of any good coming of the sort of assemblies I frequent; so I mind my own business and obey the old precept—"He that is unrighteous, let him do unrighteousness still; and he that is filthy, let him be made filthy still; and he that is righteous let him do righteousness still; and he that is holy, let him be made holy still."[3] For none of us can feel quite sure in which category the final judgment may place us; and in the meantime Miss Gaby Deslys is as much entitled to the benefit of the doubt as the Bishop of Kensington.

4. *Will not censorship always confuse voluptuousness with morbidity? What is the nature of evil in this case? And can you suppress the power for evil without suppressing also the power for good?*

A discerning correspondent who signed himself E.A.B. voiced what was to become a familiar criticism of Shaw — that he was indulging in "the passion of pure reason." Morality for Shaw, he said, was a subject for dialectics rather than for conduct; and it was E.A.B. who reduced the question in essence to: "What is the nature of evil?" Christian ethics tend to regard the body as evil. The Church, he noted, deplores sex but condones it, with much the same attitude it takes toward war. But Shaw was an "ultraChristian," obsessed like his own John Tanner by the divine Life Force. For such a person when, if ever, is sex immoral? When it is "the conscious glorification of a means to an end," E.A.B. answered himself, and the Reverend T. A. Lacey was quick to point out that this was precisely the view of Saint Augustine.

Without disputing such a definition of "immoral sex" Shaw observed how impossible the application of such a standard would be, since

> ... men have worshipped Venuses and fallen in love with Virgins. There is a voluptuous side to religious ecstasy and a religious side to voluptuous ecstasy; and the notion that one is less sacred than the other is the opportunity of the psychiatrist who seeks to discredit the saints by showing that the passion which was exalted in them was in its abuse capable also of degrading sinners.

Such religious-voluptuous emotional complexity makes it impossible to "dissect out the evil of the theatre and strike at that alone."

> One man seeing a beautiful actress will feel that she has made all common debaucheries impossible to him; another, seeing the same actress in the same part, will plunge straight into those debaucheries because he has seen her body without being able to see her soul. Destroy the actress and you rob the first man of his salvation without saving the second from the first woman he meets on the pavement.

One of the contributors slyly "confessed" that GBS had almost convinced him, at this point, that the same emotion was stimulated by the Venus de Milo as by a pornographic photo. Here again one must pause, I think, to marvel at Shaw's inability to conceive of real lewdness or pornography. While what he says about the beautiful actress must have been many times proved, it is hard to conceive of anyone renouncing all "common debaucheries" on behalf, say, of a grind-and-bump performer!

This does not necessarily invalidate Shaw's position, for it may well be impossible to "dissect out the evil" from even the most salacious performance; but it does leave a blind spot in Shaw's view of the theatre and the world, and in a sense leaves the Bishop of Kensington and the National Vigilance Committee curiously unanswered in regard to the existence of the crudest carnality.

But freethinkers in 1913 were still in the first wave of revolt from a prudery that regarded the least voluptuous art as indecent, and Shaw was more concerned with the effects of such "starvation" than with the effects of commercial sensuality.

> We have families who bring up their children in the belief that an undraped statue is an abomination; that the girl or youth who looks at a picture by Paul Veronese is corrupted for ever; that the theatre ... is the gate of hell; and that the contemplation of a figure attractively dressed or revealing more of its outline than a Chinaman's dress does is an act of the most profligate indecency. Of Chinese sex morality I must not write in the pages of *The Times*. Of the English and Scottish sex morality that is produced by this starvation and blasphemous vilification of the vital emotions I will say only this: that it is so morbid and abominable, so hatefully obsessed by the things that tempt it, so merciless in its persecution of all the divine grace which grows in the soil of our sex instincts when they are not deliberately perverted and poisoned, that if it could be imposed, as some people would impose it if they could, on the whole community for a single generation, the Bishop, even at the risk of martyrdom, would reopen the Palace Theatre with his episcopal benediction, and implore the lady to whose performances he now objects to return to the stage even at the sacrifice of the last rag of her clothing.

In spite of Shaw's lucid rhetoric it seems to have escaped many of the correspondents that he remained the strictist of moralists and a natural puritan. He once told Archibald Henderson, "I could not write the words Mr. Joyce uses: my prudish hand would refuse to form the letters." Yet he could foresee that "For all we know [such words] may be peppered freely over the pages of the lady novelists of ten years hence; and Frank Harris' autobiography may be on all the bookstalls."[4] In spite of his own distaste for blatant and vulgar sex, his position in these letters, as elsewhere, was that morality must be imposed by self-discipline and that dependence upon the censor is the easiest excuse for immorality. Both art and religion are powers for evil as well as for good, but a man must choose his own salvation or damnation.

> An evil sermon — and there are many more evil sermons than evil plays — may do frightful harm; but is the Bishop ready to put on the chains he would fasten on the playwright, and agree that no sermon should be preached unless it is first read and licensed by the Lord Chamberlain? No doubt it is easier to go to sleep than to watch and pray; that is why everyone is in favour of securing purity and virtue and decorum by paying an official to look after them. But the result is that your official, who is equally indisposed to watch and pray, takes the simple course of forbidding everything that is not customary; and, as nothing is customary except vulgarity, the result is that he kills the thing he was employed to purify and leaves the nation to get what amusement they can out of its putrefaction. Our souls are to have no adventures because adventures are dangerous.

All this was in November of 1913. Shaw, as we have noted, was about to prepare his Preface for the published version of *Androcles*. No doubt the exchange whetted his mind for the writing of that essay, which he did not complete until the end of 1915. But on the specific question of sexual morality he found that the Jesus of the Gospels was no real help. "Sex is an exceedingly subtle and complicated instinct," he was to write in the Preface; "and the mass of mankind neither know nor care much about freedom of conscience, which is what Jesus was thinking about, and are concerned almost to obsession with sex, as to which Jesus said nothing" (IV, 539).

Gaby's gyrations would no longer cause an Anglican bishop to raise an eyebrow, and lady novelists do indeed pepper their pages with Joyce's four-letter words. But the problem of pornography has by no means been resolved. And from the depths of a culture submerged in "adult entertainment" and X-rated films, some of the sentences from *The Times* exchange must seem at least as trenchant — and as controversial — as they ever could have been in terms of the London theatre of a couple of generations ago.

5

Pacifism and the Quakers

When war actually broke out, Shaw retired to Torquay for two months and devoted himself to the writing of his pamphlet, *Common Sense About the War.* The stupidity and hypocrisy of the diplomatic fumbling which led to the debacle infuriated him as no other series of human foibles in his entire lifetime. He was particularly bitter about the abject failure of the churches to maintain any shred of Jesus's peace testimony. Yet, in spite of his respect for the sanctity and dignity of life, his vegetarianism and his anti-vivisectionism, he was never a doctrinaire pacifist. Stupid as it may be, "... when war is once let loose," he wrote, "and it becomes a question of kill or be killed, there is no stopping to argue about it; one must just stand by one's neighbors and take a hand with the rest. . . ."[1] Also he preferred executions by the state to the cruelty of long imprisonment;[2] and later in his life, when he was struggling with the problems of government, he advised that those who were determined to interfere with the practice of tolerance for others should be "liquidated."[3]

Nevertheless the plight of the conscientious objector in World War I (the "Conshy" as he was then called) touched his humanitarianism. He defended the Conshies in print, and, at least on one occasion, before a court-martial board. These activities brought him into contact with the Society of Friends and began a rather stand-offish love affair with the Quakers that continued throughout the rest of his life. Indeed in his ninety-fourth and last year he wrote for the *Atlantic Monthly*:

> I ... being often challenged to denominate myself ... have ceased to reply that my nearest to an established religion is the Society of Friends, and while calling myself a Creative Evolutionist, might also call myself a Jainist Tirthankara as of eight thousand years ago.[4]

Blanche Patch recalls his reply when a member of the press inquired (perhaps facetiously) whether he had not turned Roman Catholic: "I am not a Roman Catholic. . . . If I had to be fitted into any religious denomination, the Society of Friends, who are at the opposite pole to the Roman Catholics, would have the best chance." But he goes on to repeat once again that in view of his very explicit writings on the subject there is no excuse for describing him as a member of any of the churches.[5] At one point he told S. J. Woolf that he had the same religion as that of Dean Inge: "We are both Quakers," he explained. "We don't believe in set prayers. When we want to talk with God, we use the same language that we ordinarily use, not prayers composed for us by other people, and we do not need a church to hold communion with him."[6]

A similar report is given by Stephen Winsten, his Quaker neighbor at Ayot St. Lawrence:

> One Sunday when I had just returned from a Friends Meeting, he and I talked about the Quaker faith. He said he was a Quaker by temperament but not by faith. He could not define his faith and did not want to, but the accepted mythologies did not appeal to him. He said:
>
> "What an amazing title for a religious organization: Friends! That in itself was a stroke of genius. I believe in the discipline of silence and could talk for hours about it . . ."[7]

Then he told Winsten that he believed the "spontaneous" prayer of the Quakers involved long and arduous preparation just as his own "spontaneous" speeches always did; and he further chides Friends for denying the kind of religious healing that is to be found in art and music. In nearly all his statements about Friends he shows himself to be more understanding of their historical positions than of their contemporary practices. Consequently when he predicted that the next generation would not be puzzled about classifying him as "a Quaker of sorts,"[8] he was slightly off the mark. Most Friends, a generation later, *would* question the classification, even with the limiting phrase *of sorts* appended.

He was, indeed, a powerful spokesman for many of the testimonies normally associated with the Quakers, but it is not known that he ever attended a Quaker Meeting for worship or participated in any experience of *group* mysticism which is the basis of Quaker faith and practice. Nor does it appear that he made any effort to associate himself with the British and the American Friends Service Committees, although they both emerged from the crucible of the First World War and were concerned with the same problems of conscience that enlisted his own attention to Quakerism. He was apparently satisfied to think of Friends as a composite of the seventeenth-century George Fox, a rather stereotyped image of a benevolent nineteenth-century factory owner, and a handful of his contemporaries — mostly conscientious objectors whom he did not know personally.

It was the absolutist position of the Conshies that challenged his own ambivalence in regard to pacifism. At the time of the Boer War at the turn of the century Shaw found it difficult to support W. T. Stead's Stop the War Committee. "Do you expect me solemnly to inform a listening nation that the solution of the South African problem is that the lion shall lie down with the highly-armed lamb in mutual raptures of quakerism, vegetarianism, and teetotalism?" he wrote to his fellow-Fabian, George Samuel. "Two hordes of predatory animals are fighting, after their manner, for the possession of South Africa, where neither of them has, or ever had, any business to be from the abstractly-moral, the virtuously indignant Radical, or (probably) the native point of view." But on the whole he preferred a British victory because he objected to "stray little states lying about in the way of great powers," and because he thought a British commonwealth augured better for socialism than another independent capitalist society like America.[9]

Perhaps some of the invective that he poured on the heads of the British public during the years of the 1914–1918 war and after was an expression of his own frustration at being unable to resolve in his own mind the basic conflict of *Major Barbara*. The "way of life [that] lies through the factory of death," and "the unveiling of an eternal light in the Valley of the Shadow" (III, 184) were, in 1914, no longer philosophical concepts in a play but grim realities ending in long casualty lists. At the beginning of the war, Britain had no conscription act, and was dependent on voluntary recruiting. With suppressed rage Shaw describes the patriotic young women presenting young civilians with white feathers and singing, "Oh we dont want to lose you; but we think you ought to go," which Shaw found particularly irritating in its cockney rendition.[10] But in 1916 a conscription act was passed, putting an end to the white feathers and the singing — and to a long-cherished tradition of individual liberty. The act did contain a clause (as did the United States Selective Service Act the next year) exempting men who could show a conscientious objection to war.

The path of the true Conshy, however, was not an easy one. He had to present before a local board some evidence of the sincerity of his position. And even if he succeeded in obtaining his exemption, his life as a civilian was often made miserable by the contempt and sometimes maltreatment of former friends. But for Shaw, "as far as the question was one solely of courage, the Conshy was the hero of the war."[11] And he may have envied them their ability to resolve a moral dilemma by rejecting violence *in toto*. They were Androcles, but they had no protecting lion to shield them from the emperor.

Shaw came to their rescue without ever being wholly convinced. The trouble with militarism (he was saying in 1914) is that it must keep feeding on more and more victories. Napoleon, for instance, conquered and conquered, yet had to go on fighting. "After exhausting the possible, he had to attempt the impossible and go on to Moscow. He failed: and from that moment he had better have been a Philadelphia Quaker than the victor of Marengo, Austerlitz, Jena, and Wagram."[12] (Why a *Philadelphia* Quaker it is hard to say, unless he considers this a sort of superlative degree!) In his defense of the Conshies he was somewhat devious. The one for whom he gave evidence at a court-martial was in reality a Freethinker, but realizing that a military court would not hold that a Freethinker had a conscience, he simply produced evidence that the man was a "fanatic" and left it to the court "to infer that a fanatic must be a hyperpious Quaker."[13] The ploy was apparently successful. But there were others who for one reason or another could not pass the "conscientious" test and spent years in prison. In letters to the editor Shaw called attention to their plight. In the case of the prominent pacifist, Clifford Allen, who was in poor health, he felt imprisonment was the equivalent of a death sentence, and he asked in the *Manchester Guardian* what the intention of the government was. Was he to be killed "because the public has no knowledge and the authorities no sense?"[14] Still, when Shaw read of the thirteen-year-old Brooklyn girl who had protested against collecting funds from schoolchildren to build a battleship, and suggested (like a good little Fabian) that the money would be better spent in improving child-labor conditions, Shaw sent her one of his postcards:

> The point about the factory children is well taken, but at present I think you had better have both a fleet and a factory act. There are too many rogues about for honest men (such as they are) to be quite safe without weapons.[15]

His vigorous defense of pacifists was not an endorsement of their pacifism.

There were more illusions to be shattered in World War I than there were in World War II. At least in the early years the 1914 war was a holy crusade against the savage Hun, and, for the faithful, Shaw's wartime writings ranged from confusing to downright traitorous. He was solemnly expelled from the Dramatists Club, and very nearly from the Society of Authors. Old friends refused to sit in the same room with him. Shaw bore all this with remarkable equanimity and good humor. Years later, with World War II palpably on the horizon, he recalled these experiences as he spoke by radio to America:

> The pacifist movement against war takes as its charter the ancient docu-

> ment called "The Sermon on the Mount." ... The sermon is a very moving exhortation, and it gives you one first-rate tip, which is to do good to those who despitefully use you and persecute you. I, who am a much hated man, have been doing that all my life, and I can assure you that there is no better fun; whereas revenge and resentment make life miserable and the avenger hateful.[16]

But he hastens to add that the commandment "Love one another" is a stupid refusal to accept the facts of human nature, of the same order as advising monkeys to become men, or cockatoos birds of paradise. With the same individual morality he championed in the matter of Gaby Deslys, he advised that our inability to love everyone does not give us any right to injure our fellow creatures, "however odious they may be."

The exaggerated responses to his early pronouncements on World War I did not, of course, discourage him from continuing to lecture the British and their allies right through to the Armistice and the Versailles Peace Conference. So much of his energies went into war and peace commentary, in fact, that he almost ceased to operate as a dramatist. His commentary was always eagerly seized upon by the press, for it was good copy and was bound to have wide readership. But his political advice was not considered particularly astute, and little of it was followed. He spoke of one of his utterances as having "about as much effect on the proceedings at Versailles as the buzzing of a London fly has on the meditations of a whale in Baffin's Bay."[17] In the years between the wars, as he watched the rearmament of the West and the failure of the League of Nations, he became more and more convinced that no existing form of government was adequate to stand up to the stresses inherent in our twentieth-century social structures, and that the collapse of Western civilization was threatening. At the age of eighty-one, in the same speech in which he gave partial acceptance to the Sermon on the Mount, he agreed also that "If we want the war to stop, we must all become conscientious objectors."[18]

Shaw's affinity for the stubbornness of the Conshies during World War I appears to have moved him to read the *Journal* of the Quaker founder, George Fox. I know of no specific reference that Shaw makes to Fox's *Journal*, but the picture he draws of Fox could hardly have emerged from any other source, and the words he puts into Fox's mouth are sometimes close to *Journal* quotations. The meeting of these two heretics across the years would seem almost to have been predestined.

The seventeenth-century Fox, it must be remembered, set out, singlehanded and, as he claimed, divinely inspired, to oppose traditional Protestantism and ecclesiasticism, to restore the voice of God to the individual, to revive prophecy, to set up his life and behavior as a testimony to the Inner Light. Except that Shaw made use of a highly polished brand of literary and dramatic communication instead of Fox's direct, personal, unliterary verbal assault, and except that he was master (and victim) of a humorously detached point of view as opposed to Fox's intensely personal and humorless one, they had much the same mission.

Fox was not at all typical of the sect he founded. A more typical choice might have been Fox's fascinating convert, William Penn; or the eighteenth-century reformer of English prisons, Elizabeth Fry; or the devout American tailor, John Woolman, who shamed his fellow Quakers into freeing their slaves a century before the Civil War. But for Shaw, Fox was the Quaker *par excellence*. Shortly after World War I, he told his biographer, Hesketh Pearson, that he wanted to write a play on George Fox and on a religious theme.[19] We may regret that the play never materialized, but eventually a play did appear with George Fox in it. Shaw did not get around to dramatizing Fox, however, until late in 1938, and then, in his eighty-third year, he decided, perhaps with a sense of haste, to include a number of other favorites of his from the end of the seventeenth century and to combine them into "an imagined page of history," "*In Good King Charles's Golden Days*." "Charles might have met that human prodigy, Isaac Newton," Shaw explains in his preface. "And Newton might have met that prodigy of another sort, George Fox, the founder of the morally mighty Society of Friends, vulgarly called the Quakers" (VII, 204).

In his later plays, as we shall note, Shaw broke drastically from orthodox dramatic structure. *Good King Charles* is really a long one-act play (lasting more than an hour) with a twenty-minute epilogue. The main body of the discourse takes place in the rooms of Isaac Newton. They are invaded, against the protestations of his housekeeper, by Charles II, Fox, the painter Godfrey Kneller, three of Charles's mistresses, young James (who is to become James II), and others. Shaw thought of Hogarth as the painter, but since Hogarth had not yet been born, the artist's lines were put into the mouth of Kneller. The only external action arises from the futile attempts to protect Newton's privacy and to contain the squabbling of the King's mistresses. It is principally a brilliant talk-fest on subjects ranging through politics, art, morality, and religion. The second-act epilogue is a contrasting, pleasant, homelike scene between Charles and his Queen, Catherine of Braganza. The play still awaits a full-scale professional production in the United States, though it had a brief success off-Broadway, and a successful revival at Canada's Shaw Festival at Niagara-on-the-Lake in 1981.

The play has something of the flamboyant comic-opera quality — one almost expects any of the characters to break into a Mozartian aria. The style therefore places the fiercely puritan Fox at a special disadvantage. Furthermore, though in life he was not afraid to carry his message to the great and the near-great, he was essentially rural — a shepherd boy — not accustomed to the court and "society." Nevertheless, given the circumstances and the style which Shaw has imposed, Fox's words and behavior ring true for the founder of "the great Cult of Friendship."

The time of the play is 1680 — Fox is fifty-six. He rejects the title of Mister for himself, but allows himself to be called Pastor. He says, "I am not one of those priestridden churchmen who believe God went out of business six thousand years ago when he had called the world into existence and written a book about it. We three sitting here together may have a revelation if we open our hearts and minds to it" (VII, 221). At one point the sound of a church bell sends him into a hysterical rage — over-theatrically, perhaps, but correctly illustrative of his attitude toward the "steeple houses."

Shaw had always been fascinated by the fact that the great scientific mind of Isaac Newton had been occupied seriously with making a chronology of the Bible similar to that of Bishop Ussher (a more impressive intellectual feat, Shaw noted, than his law of gravitation), and that it was this work for which he most hoped to be remembered. Newton finds no sympathy from Fox in this occupation: "The Lord does not love men that count numbers," says the Quaker. "Read Second Samuel, chapter twenty-four" (VII, 237). When Newton tries to place the beginning of the universe at the year 4004 B.C., Fox admonishes him: "Are ten million years beyond the competence of Almighty God? They are but a moment in His eyes. Four thousand years seem an eternity to a mayfly, or a mouse, or a mitered fool called an archbishop. Are we mayflies? Are we mice? Are we archbishops?" (VII, 238).

But one of the characteristics of Shavian drama is that all the characters find themselves challenged in such a way that they have an opportunity to learn something if they will. George Fox finds himself forced to treat the scandalous Nell Gwynn as a "friend" and a child of God. And he is further upset by the King (whom he addresses simply as Charles Stuart) when the suggestion is made that a playhouse, being a place where two or three are gathered together, might be as divine as Fox's meetinghouse. He must listen, too, to effete dramatic recitations from Dryden's *Aurengzebe*, which in turn precipitates a discussion on pleasure. Shaw has Fox say, perhaps for both of them, "It is not pleasure that makes life worth living. It is life that makes pleasure worth having. And what pleasure is better than the pleasure of holy living?" (VII, 266). But the most disturbing challenge to the founder of the Quakers in this brilliant company is that there are large areas of revelation common to them all. "Things come to my knowledge by the grace of God," Fox notes, "yet the same things have come to you who live a most profane life and have no sign of grace at all" (VII, 269). And at one point Shaw manages impishly to have Fox "out-Quakered." It is the artist, Kneller, who implies that his hand, while painting — while making profane objects of art — is as inspired as Fox's inner voices, that his hand is, in truth, the hand of God. When Fox is dumfounded by this, Kneller responds with a Quakerly query: "And whose hand is it if not the hand of God?" (VII, 279).

Later, when young James quotes pessimistically from Dryden ("How vain is virtue ..."), Fox takes no exception. "I have too good reason to know that it is true," he says. "But beware how you let these bold impious fellows extinguish hope in you. Their day is short; but the inner light is eternal" (VII, 264). After more than an hour of provocative talk, Newton's housekeeper manages to usher the uninvited guests into another room for a makeshift dinner. In the brief concluding scene Charles finds repose with his Portugese wife, Catherine, for whom all the troublesome questions that were argued in Act I are resolved in her Catholicism. "*In Good King Charles's Golden Days*" is not an ordinary evening in the theatre. It is rather, as one of the original reviewers commented, like sitting at a dinner party with five good talkers going at it hammer and tongs. It reveals nothing new in Shaw's religious philosophy, but it creates a memorable dramatic metaphor by placing Shaw's Life Force ideas in the mouth of the seventeenth-century founder of the Religious Society of Friends. A metaphor is an analogy, not an equation, and the similarity between Fox and Shaw, though interesting, would not stand too careful scrutiny.

6

The Struggle Against Cynicism

In terms of dramatic output the war years were unproductive. But in every other way they were hectic and significant for Shaw. Stanley Weintraub has examined these years trenchantly in his *Journey to Heartbreak*. Shaw found himself buffeted from every side, and he struck back with lectures (Fabian and other), a deluge of articles and letters to editors, and a series of short satiric "Playlets of the War," some of which purported to be recruiting posters in disguise. For as much as Shaw regarded the war as the natural outcome of foolish policies by selfish men, and as much as he disapproved of the war hysteria, he really did want Britain and her allies to win.

Nevertheless, the First World War's challenge to Western civilization caused him to reexamine both his political and religious pronouncements and to temper his meliorism with painful scepticism. He looked back over the decades that had brought Europe to the brink of self-destruction, and labeled them "The Infidel Half Century," which became the title of his Preface to *Back to Methuselah*. We are fortunate that during these dark years of searching he was at work on a play, a play unlike any other he had written, full of suffering and doubt, the record of a soul in torment.

Even before the war actually broke out, Shaw had begun to write *Heartbreak House*, "a fantasia in the Russian manner on English themes." It was to languish during the war years, to be more or less finished in 1916, but to await final completion in 1919 and uneventful productions in 1920 and 1921. Shaw had discovered the work of Anton Chekhov and was admittedly influenced by it.[1] But the Russian influence need not be over-emphasized. The play is, more than any of his others, a subjective reflection of his own troubled mind. Perhaps this is why today it is often regarded as Shaw's most contemporary play.

When Lady Utterword returns to her former home after an absence of twenty years, she finds

> the luggage lying on the steps, the servants spoilt and impossible, nobody at home to receive anybody, no regular meals, nobody ever hungry because they are always munching apples, and, what is worse, the same disorder in ideas, in talk, in feeling. [V, 66]

Nothing in this house is as it seems. The ancient sea-captain, Shotover, supports the strange household by inventing destructive devices. He is "a very clever man; but he always forgot things; and now that he is old, of course he is worse. And I must warn you that it is sometimes very hard to feel quite sure that he really forgets" (V, 69). Ellie Dunn, a romantic young girl who wanders into this menagerie, is forced into a cynical acceptance of a hypocritical world. The handsome and heroic Marcus Darnley, with whom Ellie has fallen in love, turns out to be in reality Hector Hushabye, the husband of the old captain's daughter, Hesione. Boss Mangan, a "captain of industry," owns only paper shares. Even the burglar who ostensibly tries to rob the house is not really a burglar. He makes his living by purposely being caught and throwing himself on the sympathy of his victims. And so on. These are charming, useless, disorganized people, who refuse to learn their business as Englishmen: "Navigation. Learn it and live; or leave it and be damned" (V, 177).

A thread of religious mysticism emerges here and there, but it does not constitute a dominant theme. In an argument over who is to have the last word, Lady Utterword tells Mangan that "Providence always has the last word." Lady Utterword is probably the last person from whom anyone would expect a religious reference, and Mangan reacts accordingly: "Now you're going to come religion over me. In this house a man's mind might as well be a football" (V, 130). Later Ellie's idealistic but ineffectual father (whose given name was, ironically, Mazzini) also mentions Providence, but almost casually: "Though I was brought up not to believe in anything, I often feel that there is a great deal to be said for the theory of an overruling Providence after all." But the old Captain will have none of it: "Every drunken skipper trusts to Providence. But one of the ways of Providence with drunken skippers is to run them on the rocks" (V, 176).

Near the end of the play, when a bomb falls from the sky and explodes the Captain's dynamite supply, blowing Mangan and the burglar to bits, Captain Shotover declares that "it is the hand of God" (V, 179). Another explosion has already destroyed the nearby church, and Shaw has Shotover declare, in a jarringly blatant piece of symbolism, "The Church is on the rocks, breaking up. I told [the rector] it would unless it headed for God's open sea" (V, 177–78). Far more resonant is Hector's earlier invective against the heavens: "I tell you one of two things must happen. Either out of that darkness some new creature will come to supplant us as we have supplanted the animals, or the heavens will fall in thunder and destroy us" (V, 159).

But these quasireligious references are all tangential. In this play there is no clear agent of the Life Force — no Peter Keegan or Barbara Undershaft or Lavinia — not even an Eliza Doolittle! The Life Force seems to have abandoned the people of Heartbreak House, and, almost, their creator.

The ancient Captain has, of course, moments of inspiration, but he has "sold his soul to the devil" in more ways than one. In the old days he invented the story to strike terror in the hearts of his unruly crew. But in later years, to support this idle household, he has taken to inventing instruments of death, since life-supporting inventions simply do not pay enough. In devising new weapons he has none of Andrew Undershaft's illusion of leading civilization through the very shadow of death so that it can emerge into a more hopeful day. His search for "the seventh degree of concentration" has degenerated into a dependence on rum. Nevertheless, in what Shaw was to call "the queer second wind that follows second childhood" (VII, 382), the old man has flashes of strange insight, and he remained one of Shaw's own favorites. In the little puppet play which Shaw wrote in his ninety-third year for Waldo Lanchester's puppets (*Shakes versus Shav*), Shav matches Captain Shotover with Shake's Lear.

It is never made entirely clear whether or not the Captain's fascinating daughters, Hesione Hushabye and Ariadne Utterword, are really the offspring of the Captain by the West Indian negress for whom he reputedly traded his soul to the devil. As Hector tells it, "Old Shotover sold himself to the devil in Zanzibar. The devil gave him a black witch for a wife; and these two demon daughters are their mystical progeny" (V, 156). And Hesione, at least, accepts the idea that they are "the devil's granddaughters" (V, 165).

Certainly both daughters have, in their own way, almost supernatural ability to entice men. Hesione is the warm sympathetic one, easily able to melt through the artificial surface of Boss Mangan to expose the frightened braggart-child underneath. It is her apparent wide-open friendliness that attracts young Ellie into the house to begin the play. Hesione is somehow at home in this Bohemian household. Ariadne, on the other hand, has revolted from it and chosen an ordered conventional life as the wife of a "governor of all the crown colonies in succession" (V, 66). But in flirting with her sister's husband, she tells him:

> I am a woman of the world, Hector; and I can assure you that if you will only take the trouble always to do the perfectly correct thing, and say the perfectly correct thing, you can do just what you like. [V, 97]

The daughters are strong characters, but their aim in life, if they really have any, is to live comfortably and well, and to get their own way. There is little evidence that the Life Force flows through *them*.

We never meet Ariadne's husband, Hastings, but we get the image of an efficient nineteenth-century colonial governor who keeps the "natives" under control with "a good supply of bamboo." This arrangement suits Ariadne, who would like to see her husband ruling all Britain. But the Captain dismisses Hastings as a numskull. "Any fool could govern with a stick in his hand. *I* could govern that way" (V, 165).

Hesione's husband, Hector, is a romantic pretender, courageous enough in his own right, but preferring to enact the heroes of his imagination. To Ellie, who has fallen in love with one of those images, he is simply a liar. But Hector is really escaping the wickedness of the world, and dreams of killing those who are responsible for it. He tells the Captain, "I must believe that my spark, small as it is, is divine, and that the red light over their door is hell fire" (V, 101). But he is a thoroughly ineffectual dreamer who has become Hesione's lapdog.

Occasionally Hector struggles to be free. Randall, Ariadne's devoted brother-in-law who follows her everywhere, does not even struggle. He allows himself to be "dragged about and beaten by Ariadne as a toy donkey is dragged about and beaten by a child." He is not even her lover. Hector puts it quite plainly: "She makes you her servant; and when pay-day comes round, she bilks you" (V, 156).

Ellie's father, Mazzini, began as a liberal do-gooder. He "joined societies and made speeches and wrote pamphlets" (V, 175). He kept expecting a revolution that never happened. Finally he joined forces with Mangan, even though Mangan legally robbed him, because Mangan could make a success of things and Mazzini couldn't. Hesione finds him, to her surprise, an interesting and even lovable man. But he is a man who has, in fact, long since stopped struggling.

The only other possible vehicle for the Life Force is Ellie Dunn. (The Nurse and the Burglar are more dramatic devices than characters.) One gathers that Shaw may have intended Ellie as a character who would awaken to her role in life as Barbara Undershaft did. But the play, with its almost Strindbergian dream-like evolution, did not take such an up-beat direction. Ellie's disillusionment when she discovers that her romantic dream-man, Marcus Darnley, is in reality Hesione's husband, Hector Hushabye, turns into hard-headed cynicism. The change is so abrupt that Hesione is not prepared for it. Ellie ruthlessly manipulates Mangan into a prospective marriage of convenience (he to be near Hesione, she to be near Hector). Then, sure of her new power, she blatantly throws him over and enters into a "mystical marriage" with the old Captain himself. But the Captain is no more a real person than Marcus Darnley was. She does not foresee a future of dynamic adventure, but only a dream-like happiness.

"Dream. I like you to dream," she tells the old man. "You must never be in the real world when we talk together" (V, 148). When she announces the "marriage" to the others, she says, "Yes: I, Ellie Dunn, give my broken heart and my strong sound soul to its natural captain, my spiritual husband and second father."

> *She draws the Captain's arm through hers, and pats his hand. The Captain remains fast asleep.* [V, 168]

In the end she joins those who yearn for their own destruction. This is a Death Wish, not a Life Force.

The tonality of *Heartbreak House* is a long way from the brilliantly clever assurance of *Man and Superman*. In the Preface to the earlier play, Shaw was lecturing: "When [someone] declares that art should not be didactic, all the people who have nothing to teach and all the people who dont want to learn agree with him emphatically" (II, 528). But now Shaw found himself, like one of his later post-war protagonists,[2] with no affirmations to preach.

It is frequently a mistake to try to divine the thoughts of the playwright through the words of his characters; but in this case the tenor is so pervasively sombre that one is forced to conclude (as with *Hamlet* or *Lear*) that the people of the play speak, at places, for the author. "I tell you I have often thought of this killing of human vermin," Hector tells Shotover. "Many men have thought of it. Decent men are like Daniel in the lion's den: their survival is a miracle; and they do not always survive" (V, 101). "Give me deeper darkness," the Captain cries as he goes to work on some death-dealing device. "Money is not made in the light" (V, 105). Ellie laments, "When your heart is broken, your boats are burned; nothing matters anymore" (V, 140). And the Captain chides her: "What did you expect? A Savior, eh? Are you old-fashioned enough to believe in that?" (V, 145).

Indicative of the way this play emerged with intuitive form directly intact from Shaw's imagination is the arrival of the zeppelin in the darkness overhead near the end of the last act. There has been no mention that a war is in progress, no mention of when the play is taking place. There is only a slight forewarning at the beginning of the act when Hesione thinks she hears "a sort of splendid drumming in the sky" (V, 159). Perhaps the airship *is* the war, suddenly and shockingly signaling an end to a useless and sybaritic way of life. After the first bomb has destroyed the rectory, the phony burglar and the phony businessman run for the gravel pit, and so to their doom. The Nurse and Mazzini might have gone to the cellar if there had been time. Randall can only do what Ariadne tells him to, which is to play "Keep the Home Fires Burning" on his flute. The others — Ellie, the Captain, Ariadne, Hesione, and Hector — all welcome their imminent annihilation. Hector actually lights all the lights in the house and tears down all the curtains to make the house a better target. But the zeppelin passes overhead and drops its remaining bombs elsewhere. The strangely assorted characters are left staring at the sky, futile but exhilarated. For Hesione it is "a glorious experience" and she hopes the airship will return tomorrow night. And:

> ELLIE [*radiant at the prospect*] Oh, I *hope* so. [V, 181]

This is an ending unlike anything else in Shaw. In its death-wish it comes close to the nihilism of some of the German Expressionists, who were also writing out of their disillusionment following World War I.

In spite of the sombre tonality the play evokes sporadic laughter. Even in the midst of excruciating pain, Shaw could not wholly escape from the detached view that made him a comedian. But the laughter is not the kind that is heard at a performance of *You Never Can Tell* or *Arms and the Man*. In no other play did the dramatist come so close to succumbing to the cynicism of Larry Doyle: "And all the while there goes on a horrible, senseless, mischievous laughter" (II, 910).

Perhaps that is why, in the aftermath of the war, the play was not widely appreciated. Even so, Shaw had delayed its production:

> When men are heroically dying for their country, it is not the time to shew their lovers and wives and fathers and mothers how they are being sacrificed to the blunders of boobies, the cupidity of capitalists, the ambition of conquerors, the electioneering of demagogues, the Pharisaism of patriots, the lusts and lies and rancors and blood-thirsts that love war because it opens their prison doors, and sets them in the thrones of power and popularity. For unless these things are mercilessly exposed they will hide under the mantle of the ideals on the stage just as they do in real life. . . . That is why comedy, though sorely tempted, had to be loyally silent; for the art of the dramatic poet knows no patriotism; recognizes no obligation but truth to natural history; cares not whether Germany or England perish; is ready to cry with Brynhild, "Lass' uns verderben, lachend zu grunde geh'n" sooner than deceive or be deceived; and thus becomes in time of war a greater military danger than poison, steel, or trinitrotoluene. [V, 57–58]

The Preface was not written until near the end of 1919. The paradoxes that wander through the unresolved drama still persist, but one senses the beginnings of the return of faith. How is it, he asks, that the soldiers who have been decorated for bravery in the field are continually being picked up back home for petty crimes?

> Strange that one who, sooner than do honest work, will sell his honor for a bottle of wine, a visit to the theatre, and an hour with a strange woman, all obtained by passing a worthless cheque, could yet stake his life on the most desperate chances of the battle-field! Does it not seem as if, after all, the glory of death were cheaper than the glory of life? If it is not easier to attain, why do so many more men attain it? At all events it is clear that the kingdom of the Prince of Peace has not yet become the kingdom of this world. His attempts at invasion have been resisted far more fiercely than the Kaiser's. Successful as that resistance has been, it has piled up a sort of National Debt that is not the less oppressive because we have no figures for it and do not intend to pay it. A blockade that cuts off "the grace of our Lord" is in the long run less bearable than the blockades which merely cut off raw materials; and against that blockade our Armada is impotent. [V, 43]

Shaw was sixty-three when World War I drew to a close, and he needed all the power that the Life Force could bestow on him to repel the defeatism and cynicism of the *Heartbreak House* characters. As he surveyed the ruins of the European catastrophe, he pondered on what had basically gone wrong — not merely the political and diplomatic errors, nor even the foolish attitudes which he had scored in *Common Sense About the War.* No, he needed now to probe deeper, into what is wrong with human nature itself. And does it have a cure?

Back of the disaster, and underlying our entire civilization, he found a lack of moral sense. And at the heart of this amorality Shaw seized upon the idea of Natural Selection — the doctrine of the neo-Darwinists. For if we have evolved merely through a series of accidental mutations, then it follows that both morality and the human will account for nothing, and we are condemned to drift without navigation to destruction.

Shaw could not accept such a hypothesis. In 1918, even before he had put the finishing touches on *Hearbreak House,* he became convinced that to survive mankind needed to be rescued from valueless agnosticism, and he set about creating a mythology that would support a belief in Creative Evolution. In so doing he was following the advice that his character, Undershaft, had prescribed for Major Barbara in 1906: "Well, you have made for yourself something that you call a morality or a religion or what not. It doesnt fit the facts. Well, scrap it. Scrap it and get one that does fit" (III, 170–71).

Part II

Serendipity or the Life Force? The Darwinians, Teilhard de Chardin, and Back to Methuselah

7

The Problem: Why?

The myth consisted of five plays grouped under the title of *Back to Methuselah*, begun in 1918 and completed, in amazingly short time, by 1920. He subtitled the work "A Metabiological Pentateuch," and near the end of the Preface he explains why.

> I knew that civilization needs a religion as a matter of life or death; and as the conception of Creative Evolution developed I saw that we were at last within reach of a faith which complied with the first condition of all religions that have ever taken hold of humanity: namely, that it must be, first and fundamentally, a science of metabiology. [V, 337]

In the course of the long Preface he states what he has come to believe is at the heart of the evolutionary process:

> You are alive; you want to be more alive. You want an extension of consciousness and of power. You want, consequently, additional organs, or additional uses of your existing organs: that is, additional habits. You get them because you want them badly enough to keep trying for them until they come. Nobody knows how: nobody knows why: all we know is that the thing actually takes place. [V, 273–74]

Now Darwin himself, is his *The Origin of Species* and in *The Descent of Man*, is willing to accept some degree of inheritance of habits, particularly as they affect the use or disuse of certain organs. But his principal thesis rests on what he calls natural selection or "the survival of the fittest." And this, we must recall, takes place only because in the process of self-reproduction, the copy is not always exactly the same as the original. If the change or mutation helps the organism to survive and reproduce in the struggle for existence, the mutation is called an *improvement* and it becomes a permanent part of advancement. If the mutation, as is much more likely, is not an improvement, it leaves the species no further advanced and may contribute toward its eventual extinction.

One of the difficulties of Darwin's contemporaries in accepting this view had to do with *time*. It then seemed inconceivable that the world had existed long enough to accommodate the slow minute changes and the requisite number of generations to result in so complex a biological terminus as man. But in the century following *The Origin of Species* geologists have extended the formation of the planet back to ten or eleven billion years, and the origins of life as far as three and a half billion. So Darwin's successors in the present century could proclaim that time was no longer a problem in accepting a strict Darwinian line of human emergence. The evidence was there, in the rocks, in the fossils, in the myriad surviving life forms on different parts of the planet.

These later Darwinians, of whom I am selecting Sir Julian Huxley as a leading representative, went much farther than Darwin in their support of natural selection. "So far as we now know," writes Huxley in about 1952, "not only is natural selection inevitable, not only is it *an* effective agency of evolution, but it is *the* only effective agency of evolution." Since he states the argument against the existence of a Life Force as dogmatically as anyone, let me quote him more fully:

> With the knowledge that has been amassed since Darwin's time, it is no longer possible to believe that evolution is brought about through the so-called inheritance of acquired characters — the direct effects of use or disuse of organs, or of changes in the environment; or by the conscious or unconscious will of organisms; or through the operation of some mysterious vital force; or by any other inherent tendency. What this means, in the technical terms of biology, is that all the theories lumped together under the heads of orthogenesis and Lamarckism are invalidated, including Lysenko's Michurinism.... They are *out*: they are no longer consistent with the facts. Indeed, in the light of modern discoveries, they no longer deserve to be called scientific theories, but can be seen as speculations without due basis of reality, or old superstitions disguised in modern dress. They were natural enough in their time, when we were still ignorant of the mechanism of heredity; but they have now only a historical interest.[1]

Although this sounds like Huxley's final word on the subject, it is not. Beneath the surface of scientific certainty, some later doubts become apparent. Huxley, as we shall see, could have been thinking only of *prehuman* evolution when he wrote the above passage. To maintain that mind and human consciousness resulted from the serendipitous mutation of cells stretches the imagination considerably farther. Sir Julian is not unaware that such a proposition taxes ordinary common sense. He confesses in a later essay,

> What is remarkable, it seems to me, is that the blind and automatic forces of mutation and selection, operating through competition and focused immediately on mere survival, should have resulted in anything that merits the name of advance or progress.[2]

Nevertheless Huxley maintained his official stance as a Natural Selectionist all his life and publicly argued against any who disagreed. In response to one of his scientific radio broadcasts in 1942, Shaw wrote him a postcard declaring that "Biology is in a bad way. The Laboratory mind is more degenerative than malaria. The descent from Huxley, Darwin, and Spencer — broken by Butler, Bergson, and Back to Methuselah — to the simpleton Pavlov is a precipitous *dégringolade* [tumble]. . . ."[3] Huxley dismissed the comment as emotional and unscientific, and seven years later, in the continuing controversy, wrote that Shaw

> here reveals his customary incomprehension of the nature and methods of science. . . . Even if neo-Darwinism did necessarily or usually encourage a fatalistic philosophy, which it does not, this would not make it untrue, or condone the state's vetoing it to teachers, research workers, or the general public. Nor does Lamarckism become fact because Mr. Shaw and the USSR Academy of Sciences feel that it would be nice if it were true.[4]

This was written after Huxley's second visit to the Soviet Union in 1945, when he had conferred with Soviet scientists, particularly Lysenko, a follower of Michurin, who claimed that the effects of grafting and other treatments of crops could be passed on to later generations. The Soviet hierarchy supported this position, not only because it promised prompt improvement of crops, but because it implied that good habits and attitudes could be transmitted genetically to help build the true communist state. Interestingly, Huxley's earlier visit to Russia in 1931 produced a largely laudatory report, *A Scientist Among the Soviets* (Chatto & Windus, London, 1932), but this was before the Lysenko controversy had erupted. After hearing Lysenko lecture, Huxley had nothing but contempt for him as a scientist, and lumped him with all the other opponents of natural selection:

> Samuel Butler, Bernard Shaw, and Lysenko may assert that evolution without the inheritance of acquired characteristics is unthinkable, but the facts proclaim the contrary.[5]

The real challenge to Huxley's dogmatism came with his discovery of the work of a paleontologist who was also a Jesuit priest, Pierre Teilhard de

Chardin. It was a challenge he eventually accepted by writing the Introduction to the posthumous English edition of Teilhard's principal work in 1958, *The Phenomenon of Man*.

The evidence and reasoning of Père Teilhard were of a different order from Shaw's, Butler's, and Lysenko's. Huxley first met Teilhard in Paris the year after the Lysenko lecture and discussion, and immediately realized that "he and I were on the same quest, and had been pursuing parallel roads ever since we were young men in our twenties."[6] (The priest, however, was six years his senior.) Huxley found him to be widely traveled and highly sophisticated in research. Teilhard had worked at a number of paleolithic sites, particularly in China, where he spent most of his middle years. His view of evolution began not with the beginnings of life, nor even with the primal earth, but with the formation of the universe. He wrote about twenty books and countless articles, but his views were in conflict with his order and with the church, so that he did not live to see any of his major works in print. He was not even allowed to give public lectures in his later years, and was denied permission, the year before his death, to travel from New York to Paris to attend a scientific congress.[7] All this was because Teilhard found evolution to be all of a piece, and failed to identify a particular moment of divine interference that might be called "creation." His view did not in any literal sense allow for man's fall and his subsequent redemption. The concepts of sin and grace were not a part of his evolutionary pattern. Teilhard could have avoided the suppression of his works by withdrawing from the Society of Jesus, but he regarded himself as a faithful child of the Church and bore his frustration with remarkable patience. He hoped his work would provide a new view of Christianity, one that would be compatible with twentieth-century science.

Huxley had maintained, as early as 1923 in his *Essays of a Biologist*, that "... not only living matter, but all matter, is associated with something of the same general description as mind in higher animals."[8] Furthermore the presence of mind and consciousness in the human animal has superseded the blind and non-conscious forces of nature, and must direct the course of evolution in the future.[9] His outspoken claim, therefore, that natural selection was the only possible agency of evolution would have to be restricted to the period before mind and consciousness became controlling factors — presumably before the appearance of man. However, Huxley's acceptance of these elements in all matter leaves the transition from natural selection to conscious direction ambiguous. And Teilhard's arguments pushed him still further. Though Huxley maintained to the end that the breakthrough into human consciousness was "brought about by the automatic mechanism of natural selection and not by any conscious effort on [man's] own part,"[10] he gives Teilhard credit for forcing scientists to see the spiritual implications of their scientific knowledge. On the one hand, "the religiously-minded can no longer turn their backs upon the natural world;" but on the other, "the materialistically-minded [cannot] deny importance to spiritual experience and religious feeling."[11] In a sense, Huxley had as much difficulty in maintaining his materialistic agnosticism in the face of Teilhard's arguments as Teilhard had in maintaining his orthodox posture with the church in the face of the natural phenomena he was investigating. Huxley, however, as the leading British humanist of his time, was not subject to censorhip by his superiors, as the Jesuit scientist was.

It is unlikely that Shaw ever heard of Teilhard de Chardin. The first English edition of *The Phenomenon of Man* did not appear until 1958, eight years after Shaw's death. There is no good reason why Teilhard should not have heard of Shaw, but I know of no evidence that he had; and Daniel J. Leary in his seminal comparative study of the two men and their philosophies makes no mention of any direct reference of the priest to the playwright.[12] Teilhard as a scientist would certainly have paid attention to any new data available on the subject of evolution, but the philosophical ruminations of an aging Irish dramatist might well have escaped him. Nevertheless both men, as Leary observes, "were disturbed by the growing rift between science and religion, matter and spirit, and ... each attempted to reconcile this Cartesian dichotomy by projecting a vision that encompasses time and space from 'in the beginning' to 'as far as thought can reach.'"[13]

Shaw, the eldest of the three, was three years old at the publication of *The Origin of Species*. As an adult, he rejected the "neo-Darwinism" of Sir Julian's grandfather, Thomas Huxley, preferring the less scientific evolutionary ideas of J. B. Lamarck. Why did Shaw find the Darwinian view philosophically unacceptable and esthetically distasteful? Not, certainly, because he found it in conflict with any orthodox theology. The inadequacy of all existing religions was, as we have seen, a recurring theme in his speeches and writings from about 1906 onwards. With Samuel Butler he rejected the purely mechanistic interpretation of evolution because, in Butler's words, it "banished mind from the universe" (V, 300). While Shaw did not dispute the evidence of natural selection — which, as we have noted, he preferred to call *circumstantial* selection — he found it the "way of hunger, death, stupidity, delusion, chance, and bare survival"; whereas he found Lamarck's way "the way of life, will, aspiration, and achievement" (V, 249).

For some reason the argument came to center on the long neck of the giraffe, which was only one of the many examples through which Darwin illustrated a radical change of species.[14] The nascent giraffes (Darwin assumes) were a type of cattle adapted to browsing on food from trees. When the population of these animals could no longer be accommodated by the food supply, only those individuals which were the highest browsers could survive and breed. Thus the circumstance of a slightly longer neck passed along to succeeding generations; and as the competition for survival became fiercer over a thousand generations the survivors were marked by longer and longer necks. But such an explanation had for Shaw "a hideous fatalism about it, a ghastly and damnable reduction of beauty and intelligence" (V, 294). The Lamarckian explanation was simply that those individuals survived and bred that had the greatest will to reach a little higher.

Huxley and the later Darwinians would, of course, dismiss such an idea as romantic. But Teilhard de Chardin's assumption of a *noosphere* (see chapter 9) as part of the evolutionary pattern gave them, at least, some slight pause. I wish now to reexamine Shaw's "unscientific" position in the light of Teilhard's vision. Though I shall restrict the arguments, in the main, to those offered by the three principal participants in the first half of the century, it must not be supposed that the controversy has subsided. The debate over the nature of the evolutionary process has been probably the most wide-ranging in the entire history of science, and it has not abated. Just ten years ago (1971), Jacques Monod published his *Chance and Necessity,* which carries forward the mechanistic arguments of Julian Huxley, supported now by recent discoveries in cellular chemistry. Mendel's defining of the gene as the bearer of heredity, and Watson's and Crick's evidence that DNA is the basis of the gene's invariant replication constitute, for Monod, "the most important discoveries ever made in biology."[15] Any changes are random and accidental. He goes on to echo Huxley:

> . . . it necessarily follows that chance *alone* is at the source of every innovation, of all creation in the biosphere. Pure chance, absolutely free but blind, at the very root of the stupendous edifice of evolution: this central concept of modern biology is no longer one among other possible or even conceivable hypotheses. It is today the *sole* conceivable hypothesis, the only one that squares with observed and tested fact. And nothing warrants the supposition — or even the hope — that on this score our position is likely ever to be revised.[16]

Still, again and again, Monod is forced to use personifications to explain this accidential behavior. He uses Francois Jacob's image that it is the "dream" of every cell to become two cells. Such inadvertent slips into "vitalism" must have amused Arthur Koestler, who, in the same year, was maintaining:

> Darwinian selection operating on chance mutations is doubtless a part of the evolutionary picture, but it cannot be the whole picture, and probably not even a very important part of it. There must be other principles and forces at work on the vast canvas of evolutionary phenomena.[17]

Even the old argument about evolutionary time has been revived as it becomes more and more obvious that we have underestimated not only the duration of evolution, but also the complexity of what has evolved. Dr. Peter Gibbs, Professor of Physics at the University of Utah, has pointed out that the number of seconds that have elapsed since the beginnings of life on this planet more than three billion years ago would be utterly insufficient for Nature to try out, at the rate of one per second, all the possible sequences of a single DNA molecule.[18]

The arguments have shifted ground in the intervening decades, as new discoveries continue to be made in the fields of archeology, anthropology, geology, and psychology — all of them relating in some way to our understanding of evolution. But for Koestler and those of similar temperament the later evidence serves only to refine the basic arguments. It does not dispel the mystery of motivation. For them, as for Shaw, the central challenge to Darwinism is not so much *how* evolution happens, as *why* it happens.

8

Lamarck vs. Darwin

If Shaw had had Teilhard as a primary scientific reference, he would not have had to rest his premise on so shaky a foundation as that provided by Lamarck. Jean-Baptiste Lamarck (1744–1829) published his principal work, *Zoological Philosophy* in 1809. In it he surveys the animal kingdom, looking back from nature's most complex creature, man, to the most simple fauna that were then known. He did accept the heretical idea of evolution, as opposed to the Biblical creation of separate species, but his notion of evolution was that of a linear progression from the simple to the complex. Thus every animal had its place along a single line of ascent. Where there were obvious gaps, he assumed that there were animals yet to be discovered. He did not think that any had become extinct. He accounted for the phenomenon of evolution in the following way: Each animal born tends to be slightly more complex than its parent. This tendency *plus* changes brought on by environment, *plus* the inheritance of acquired characteristics ("use inheritance") produced the long slow upward climb culminating in humanity. He believed also in some subtle fluid that carried the force of life with it. This fluid was spread everywhere over the surface of the earth, and under propitious conditions generated life. As to religion, he appears to be an eighteenth-century deist rather than an orthodox churchman, accepting a First Cause, and referring to the Supreme Author of all things.[1]

Clearly Shaw hopes that his references to Lamarck will bolster his ideas with the authority of a scientific name, in spite of the fact that there is little in Lamarck that Shaw, or any other turn-of-the-century thinker, could accept on a scientific basis. But Lamarck does leave room in his scheme of things for a certain amount of internal motivation in the evolutionary process, and it is this one element, rather than Lamarck's general plan of evolution, that later writers labeled "Lamarckian." The term came to mean "evolution through inherited characteristics" as opposed to natural selection. Shaw seized on it as a weighty adjective to oppose "Darwinian" or "neo-Darwinian."

The distinction is a useful one, and it is interesting to note that Teilhard also makes use of it on a number of occasions. However, characteristically, Teilhard tries to unify the opposing concepts:

> Properly understood the "anti-chance" of the Neo-Lamarckian is not the mere negation of Darwinian chance. On the contrary it appears as its utilization. There is a functional complementariness between the two factors; we could call it "symbiosis."

Teilhard believes, as Shaw does, that there is an inner force at work in evolution. In the lower forms of life "an essential part is left to the Darwinian play of external forces and to chance ... but strokes of chance which are recognized and grasped — that is to say, psychically selected." Evolution therefore becomes more Lamarckian as it becomes more complex and more conscious.

> On the one hand is the Lamarckian zone of very big complexes (above all, man) in which anti-chance can be seen to dominate; on the other hand the Darwinian zone of small complexes, lower forms of life, in which anti-chance is so swamped by chance that it can only be appreciated by reasoning and conjecture, that is to say, indirectly.[2]

All variety of evolutionists agree that the direction of evolution is from simplicity to complexity. The later scientists, including Teilhard, reject the single evolving line of Lamarck, and give us a picture of endless struggle, of thousands of false starts and paths to inevitable extinction for every tentative breakthrough; of tiny adaptations emerging through the course of thousands of generations. "Natura non facit saltum," Darwin quotes — Nature does not make a leap, but proceeds with inevitable gradualness (though there may be exceptions even to that). However, if Nature has proceeded over eons of time purely by chance, and if we are now suddenly, in these last few millenia, called upon to take charge of the process, in what direction are we to go? Surely the series of accidents that has given us consciousness and the power of reflection can furnish no moral guidelines beyond the three essential precepts that have brought us to this threshold: survive, adapt, propagate — and these would properly constitute a law of the jungle. But if there is a Life Force pushing (or, as Teilhard prefers to think, pulling) us — pulling us toward an "Omega point"; pulling us in a struggle that is by no means wholly predetermined — then we animals who evolved a mode of conscious thought and independent behavior may have important work to do in the world.

9

The Life Force, the Noosphere, and the New Religions

In his conception of the evolutionary process, Teilhard de Chardin starts at the beginning — or as close as he can get to it. We are by sidereal measurements inhabitants on a speck of dust revolving about an undistinguished star on the edge of the milky way, itself one of innumerable galaxies. But the way of evolution, as we have already noted, is from the simple to the complex, and the path toward complexity leads from the outer vastness of space towards our own speck of dust (and, presumably, to countless other such specks elsewhere).

Most of the universe consists of atomic particles, which form the simplest of elements, hydrogen, and gather gravitationally into stars, which produce far more complex chemical combinations. But the higher complexity is reserved for the planets, particularly planets which eventually support life. With life comes the development of mind and consciousness, and finally, in the human stage, what Teilhard calls "reflection" or "co-reflection" — the consciousness of consciousness. Teilhard puts the process in its simplest form in a 1953 essay, "The God of Evolution":

> a. First, at the very bottom, and in large numbers, we have relatively simple particles (corpuscles), which are still (at least apparently) *unconscious*: Pre-life.
> b. Next, following the emergence of life, and in relatively small numbers, we have beings that are *simply conscious.*
> c. And now (right now) we have beings that have suddenly become *conscious of becoming every day a little more conscious* as a result of "co-reflection."[1]

As to the last two stages, we are, of course, restricted to observations on our own planet. Nevertheless, Teilhard would insist that there is no part of the evolutionary process that can be separated from any other part. We are part of a universal pattern, just as we are part of a terrestrial one. As a paleontologist (and something of a biologist and geologist as well) he studied the earth carefully for continuing evidence of the growing complexity of the evolutionary pattern. He noted that the earth was composed of a series of concentric spheres: "the barysphere, central and metallic, surrounded by the rocky lithosphere that in turn is surrounded by the fluid layers of the hydrosphere and the atmosphere." Eduard Suess had added the biosphere, "an envelope as definitely universal as the other 'spheres' and even more definitely individualised than them. For . . . it forms a single piece of the very tissue of the genetic relations which delineate the tree of life."[2]

To these spheres Teilhard now adds another: the *noosphere* — a "thinking layer," coined from the Greek verb νοέω 'to think or reflect'. With the spread of a thinking and reflecting animal over the surface of the earth, it "gets a new skin." With this concept Teilhard implies that thought (consciousness, reflection) is not merely a process of individual minds, but a collective phenomenon involving all humanity, aided by the means of communication man has devised — language, art, music, printing, radio, television.

There is nothing particularly "anti-Darwinian" about any of this. It gives to evolution a philosophical and metaphysical thrust that Darwin and his followers were not quite willing to take, though one fancies that Darwin himself would have been quite fascinated with the idea of a noosphere! But there is something that runs through all of Teilhard's thinking that I don't think is present in the thinking of the Darwinians, or at least is not clearly identifiable. It is the sense that the whole is always present in all of the parts. In the smallest particle or electrical charge in interstellar space there is present the seed or the spirit of the entire evolutionary process, that which we already know and that which is yet to come. The conception is equivalent to that expressed in the Hindu Scripture, *Srimad Bhagavatam:* "Creation is only the projection into form of that which already exists." This is a kind of monism which goes much further than Huxley's willingness to concede a quality of "mind" to all matter. Whatever constitutes the noosphere, which nourishes the highest and most refined kind of existence of which life is so far capable, is and has always been present throughout the universe. And because Teilhard is a deeply religious man, the notion is inescapable that he conceives this presence to be an emanation of the Godhead.[3]

To support this view, early in *The Phenomenon of Man* Teilhard devotes a chapter to "The Within of Things." Although as a scientist he is quite used to looking at matter objectively, "from without," there are too many qualities of existence that cannot be wholly accounted for in this way. We accept the *within* of ourselves, but for the physicist or the biologist it is considered unscientific to depart from the observable *without.* The fact, however, that we recognize an "interior" in one form of life is enough for Teilhard to "ensure that, in one degree or another, this 'interior' should obtrude itself as existing everywhere in nature from all time."

> Since the stuff of the universe has an inner aspect at one point of itself, there is necessarily a *double aspect to its structure,* that is to say in every region of space and time — the same way, for instance, as it is granular: *co-extensive with their Without, there is a Within to things.*[4]

Obviously the *within* of higher forms of life is more evident than it is in lower forms or in inanimate matter. Consequently Teilhard conceives of evolution as proceeding from "a very large number of very simple material elements (that is to say with a very poor *within*)" to a state "defined by a smaller number of very complex groupings (that is to say, with a much richer *within*)."[5]

This *within* of things is for Teilhard the source of spiritual energy, what Shaw called the Life Force, and what Henri Bergson called the *élan vital.* Both men acknowledged a debt to Bergson, but in neither case should the debt be overemphasized. Teilhard refers to Bergson only in his later work, and though the references are generally sympathetic, he occasionally begs to differ. He welcomes Bergson's emphasis on intuition, for example, but cautions that we must not overlook the human brain as "the most 'centro-complex' organ yet achieved to our knowledge in the universe."[6] Though challenged by Bergson's main thesis, he could not accept a purely intuitional non-rational approach to the world. And he wanted to change Bergson's phrase, *vis a tergo* — "a push from behind" — as a description of the evolutionary force to one which would imply a *pull* toward our destiny.[7]

Shaw, for his part, had his John Tanner talking about the Life Force in *Man and Superman* in 1903. Bergson's *Creative Evolution* was not published until 1907 and its English translation did not appear until 1911. Shaw's "Life Force" was not, therefore, an adaptation of Bergson's term, "*élan vital,*" as is often implied. Shaw, in fact, does not mention Bergson by name until 1912, and then merely in a footnote reference added in the second edition of *The Quintessence of Ibsenism.* We have noted how, in the meantime, both on the platform and in print, he continued to preach the driving force back of evolution.[8] By the time he wrote *Back to Methuselah,* he had, of course, acknowledged Bergson; but it is certainly an exaggeration on the part of Bertrand Russell to call the play "pure Bergsonism."[9] In the first place Bergson's conception of time would not permit such imaginative forays into the future. It is futile to think "that the future can be read in the present," he says.[10] And he is less sanguine about the inheritance of acquired characteristics than Shaw is.[11]

Shaw never looked back to the molecular origins of the Life Force, but concerned himself with its role in human behavior and human destiny. He was committed neither to a theological position, as Teilhard was, nor to the humanist agnosticism of Huxley. He anticipated, and later endorsed, the Bergsonian view that the evolutionary process was a struggle toward perfection which could not be accounted for by accidential circumstances. The motivating force had to be a part of human consciousness. Besides his own term, the Life Force, he was willing to use Bergson's *élan vital*, or the Divine Spark, or even Divine Providence. We have seen that sometimes, in his speeches, he would go so far as to call it "God." But in these cases he tried to make sure that the outworn mythology associated with that word did not get in the way.

In general terms, then, if not in particulars, Bergson's view is supportive of both Shaw and Teilhard:

> A hereditary change in a definite direction . . . must certainly be related to some sort of effort, but to an effort of far greater depth than the individual effort . . . inherent in the germs they bear rather than in their substance alone, an effort thereby assured of being passed on to their descendants.[12]

Interestingly, Shaw, Teilhard, and Huxley all proposed a "religion of the future," and all three conceived it as a religion of evolution. For Teilhard it would have to be a form of Christianity *in extenso*. "[It] cannot fail to appear before long: a new mysticism, the germ of which (as happens when anything is born) must be recognizable somewhere in our environment, *here and now*." In the new Christianity, Christ would have to be granted two new attributes: He could no longer be restricted to the redemption of our own little planet; and he would have to be conceived of as the Omega-point of evolution (a concept which we shall return to shortly).[13]

Huxley would call his religion "Evolutionary Humanism," and would make clear from the outset that it has nothing to do with a belief in the supernatural. People need a "background of reverence and awe" for their beliefs. They also have an innate need for ritual, and they often require a strong reference point for acceptable social behavior. Humanity as a whole must develop a common religion to provide common goals and a productive future. Again and again Huxley feels the need for quasi-religious terminology in his essays (e.g., "transcendence," "fulfilment"), but he carefully guards against any indication of the existence of a higher force or a mystical presence. He feels strongly attracted to Teilhard and sympathizes with his "spiritual travail," but in the end he finds it "impossible to follow him all the way in his gallant attempt to reconcile the supernatural elements in Christianity with the facts and implications of evolution."[14]

Shaw, in an address entitled "The Religion of the Future" delivered to the Cambridge Heretics in 1911, also urged the abandonment of Christianity, but preached the gospel of an un-Huxleyan mysticism that would take the Life Force into account. Taking the risk, once again, of referring to the Life Force as God, he told the Heretics:

> We are all experiments in the direction of making God. What God is doing is making himself, getting from being a mere powerless will or force. This force has implanted into our minds the ideal of God. We are not very successful attempts at God so far, but I believe that if we can drive into the heads of men the full consciousness of moral responsibility that comes to men with the knowledge that there never will be a God unless we make one — that we are the instruments through which that ideal is trying to make itself a reality — we can work toward that ideal until we get to be supermen, and then super-supermen, and then a world of organisms who have achieved and realized God.[15]

10

Our Place in the Present Universe

In "In the Beginning" — Part I of *Back to Methuselah* — Shaw follows the accepted precept that the path of evolution is from the simple to the more complex. Drama, by its very nature, can concern itself only with the legend of *human* evolution, and Shaw conceives the beginning point of that process to be at that moment when man accepts responsibility for his own future, for the rearing of children, and accepts also the inevitability of his own death. The simple unexamined life in the Garden of Eden was anti-evolutionary until it was challenged.

Back in 1968 when Robert Kennedy was running for the Democratic nomination for President, he was accustomed to closing his whistle-stop speeches with the quotation: "You see things; and you say 'Why?' But I dream things that never were; and I say 'Why not?'" He customarily prefaced the quotation with, "As Bernard Shaw said ..." He perhaps thought this more politic than to say, "As the Serpent said ..." For it is the Serpent who is the motivating force in the first scene of the play. The Serpent, whispering to Eve on behalf of the Life Force, plants the seed of discontent, and so begins civilization. Whatever nostalgia the Garden might have had for Adam and Eve in retrospect, the decision to take responsibility for their own world was a noble one (though Adam took it reluctantly). And so the Shaw myth, unlike its Biblical source, presents the Serpent as the agent of Lilith and the Life Force, and the expulsion from the Garden as the launching of the human experiment. In Act II, "a few centuries later," the problems and conflicts arising from human aggregation have already become challenging. Men and women have found it necessary to divide labor and become specialists in agriculture, spinning, rearing children, making war, and making art.

The neo-Darwinians have come to accept at least this part of the Shaw myth: Though pre-human evolution may have proceeded purely by chance, the future is in the hands of humans. "Nature —" writes Julian Huxley, "if by that we mean blind and non-conscious forces — has marvelously produced man and consciousness; they must carry on the task to new results which she alone can never reach."[1] Furthermore, we must now regard the aggregation of social units as a part of evolution. Within our complex civilization, the process of evolution works through whole communities, not merely through individuals. "There is what may be called an ecological metabolism of the whole community, just as there is a physiological metabolism of the individual species."[2]

This is in remarkable agreement with Teilhard, who carries the idea much further. Having evolved organisms that are conscious of their own consciousness, evolution is now in the process of uniting them into a still more complex organism — a society. Indeed Teilhard sees no viable path open to us except the "planetization" of mankind. He is aware that many will see this as "a crude phenomenon of mechanization or senescence which will end by dehumanizing us," but he maintains on the contrary that it is "an effect of biological super-arrangement destined to ultra-personalize us."[3] Our inability to regard cooperation and collectivization as virtues may yet doom the entire human experiment to failure; for now that we have taken control of our own development, it is by no means certain that we shall not come to disaster and even extinction as so many other species have done.

> . . . the possibility has to be faced of Mankind falling suddenly out of love with its own destiny. This disenchantment would be conceivable, and indeed inevitable, if as a result of growing reflection we came to believe that our end could only be collective death in an hermetically sealed world. Clearly in the face of so appalling a discovery the psychic mechanism of evolution would come to a sudden stop, undermined and shattered in its very substance, despite all the violent tuggings of the chain of planetary in-folding.[4]

Shaw, who was a Fabian Socialist long before he was a Creative Evolutionist, was more concerned with the social order than he was with biology. His vigorous reaction against Darwinism in the Preface that he titled "The Infidel Half Century" was triggered by the rise of opportunism in Western civilization which he felt was linked with the acceptance of natural selection as a philosophical tenet. The belief that the world and all it contains evolved from a series of accidents destroyed not only the old conception of God, but the old morality as well, and left the world at the mercy of laissez-faire materialism. Whereas everybody felt enormously relieved to be rid of the old burdens, they had not yet come to realize that they had been left nothing but a void. Shaw, then, like Teilhard, was concerned with our place in the universe and our future, but he sees the failure of the human experiment as a more remorseless threat than Teilhard does:

> The power that produced Man when the monkey was not up to the mark, can produce a higher creature than Man if Man does not come up to the mark. What it means is that if Man is to be saved, Man must save himself. There seems to be no compelling reason why he should be saved. He is by no means an ideal creature. At his present best many of his ways are so unpleasant that they are unmentionable in polite society, and so painful that he is compelled to pretend that pain is often a good. Nature holds no brief for the human experiment: it must stand or fall by its results. If Man will not serve, Nature will try another experiment. [V, 267]

The price, therefore, that men and women must pay for their powers of reflection and independent action is the responsibility for taking charge of the evolutionary process. Shaw is, of course, anti-determinist; but it is hard to deduce just how determinist either Teilhard or Huxley is. It is obvious, for example, that both of them see certain developments, such as the "planetization" of mankind, as inevitable. We can speed up the process of Nature and encourage such developments, or we can fly in the face of them and be destroyed. One way of taking nature in hand and speeding up evolution is through eugenics, by which men and women can achieve within a few generations results that Nature would require a thousand generations of trial and error to accomplish. Teilhard's approach to this subject was thoroughly non-Catholic,[5] and completely in agreement with Huxley and the humanists. Teilhard also favored helping Nature to take the next step in the direction of uniting the human race into a single complex organism by encouraging the spread of communications, the formation of more all-inclusive governments, and, one gathers, multi-national corporations.

A number of stumbling blocks in the way of our taking charge of our own destiny have become more apparent in the meantime. Teilhard and Shaw lived barely long enough to react to the atomic bomb and the beginnings of nuclear power. Teilhard was challenged and sobered by the enormity of the power now in men's hands.[6] Shaw found the bomb an atrocity and a threat to the future (VII, 428) — "vivisectional amorality applied to ourselves more ruthlessly than to the dogs."[7] But, at their deaths (Shaw in 1950 and Teilhard in 1955), the nuclear age was really just beginning, and they had little chance to ponder its implications. Neither did the ecological threats to survival, of which we have become forcibly more and more conscious through the sixties and seventies, deeply disturb their thinking, or affect their predications of the future. Teilhard incorrectly assumed that "Where physical energy and even inorganic substances are concerned, science can foresee and indeed already possesses inexhaustible substitutes for coal, petroleum, and certain metals,"[8] but he was concerned with food production and the population explosion. Shaw, in more general terms, continued to worry about "whether the human animal, as he exists at present, is capable of solving the social problems raised by his own aggregation, or, as he calls it, his civilization" (V, 260). Sir Julian Huxley lived until 1975 and seemed well aware that we are in the midst of a revolutionary phase of history, where old political and economic systems are falling apart, and new systems are struggling to evolve. He gave considerable attention to the population problem and to eugenics,[9] and lectured on health, education, and public planning. But he said surprisingly little about pollution, the depletion of our resources, and the threat of nuclear destruction.

The substantive difference in these three views appears, then, not to be so much about man's present status as about the nature of evolution in its pre-human stages. And since there seems to be essential agreement that the human race is now in a position to direct its own evolutionary development, we must ask whether it is really significant that disagreement remains about the nature of pre-human development. The answer is that the significance of the disagreement is profound. The apparent convergence of views on man's present place in the universe is deceptive. If the *direction* of evolution has been set by the chance factors which Darwin labeled "improvement," then the jungle values — survive, adapt, propagate — underlie all future development. What we have come to consider humanistic or spiritual values becomes a thin veneer disguising the brutal competition that must always remain the basic pattern in the direction Nature has set for us. There is nothing in this pattern to suggest, for example, that compassion is of a higher value than cruelty, or that honesty is superior to duplicity. Indeed in the competition to survive, adapt, and propagate, duplicity and cruelty may be far better values than compassion and honesty. Our civilization is constantly faced with these choices, and whenever we opt for cruelty or deception we justify our decision by claiming that we have to be "realistic" — an admission that we consider the *real* values to be those of Darwinian improvement, and not those that we profess in religion or politics. If, on the other hand, we could assume that a sense of fair play, of cooperation, of integrity, of the Pauline concept of love — if we could assume that such qualities were *also* inherent in the evolutionary process (that they were, in effect, part of the process of adaptation), then it would be possible to construct a code of behavior that would rest on honest values rather than on hypocritical ones, and a civilization that aimed at nobility rather than mere pragmatism.

(It is worth nothing, parenthetically, that just as the argument on the nature of physical mutations has continued into the present, so has the argument about psychological development. Konrad Lorenz and his followers find humanity endowed with a legacy of territoriality and aggression that must make us inescapably violent and belligerent. Richard E. Leakey and Roger Lewis in their *Origins* find the Lorenz arguments "not relevant to human behavior," and read from the clues given that it is much more likely that "we are a cooperative rather than an aggressive animal.")[10]

All the humanists, including Huxley, hope for a world in which what are generally known as the Judeo-Christian virtues predominate.

> I believe [writes Huxley] that there exists a scale or hierarchy of values, ranging from the simple physical comforts up to the highest satisfactions of love, esthetic enjoyment, intellect, creative achievement, virtue. I do not believe these are absolute or transcendental in any sense of being vouchsafed by some external power or divinity; they are the product of human nature interacting with the outer world.[11]

Thus the humanists' hope for the future rests on the assumption that human intelligence has evolved so far that it can reject some of the more primitive impulses of the past, and can now make the more sophisticated choices necessary for an ongoing civilization, based on cooperative principles and technological advancement. Nature has, almost miraculously, given humanity this opportunity. But, say the Darwinians, there is no divinity in the miracle. The miracle was an accident.

Teilhard, as a Christian, accepts the spirit of Christ as being present throughout the universe, and accepts the continuing struggle between the forces of good and evil, between the creative and destructive. Evolution — God — is pulling all existence toward an Omega-point, a direction and end that is far in advance of any Darwinian improvement, and, in fact, of our own comprehension.

Only Shaw senses the void between these two views. Though he is essentially in sympathy with the mysticism that Teilhard represents, he cannot believe that Christianity will be able to remake itself into a viable "religion of evolution." But to leave the field to the neo-Darwinians would be to abandon civilization to a valueless universe, where "glory, beauty, truth, knowledge, virtue and abiding love" would be unattainable. These are the qualities he listed in 1929 as having been achieved "only on paper."[12] He had therefore been forced to put his faith in a religion of the future.

11

The Future of the Human Race

Shaw alone insisted that the necessary changes — biological, psychological, and cultural — could be made possible by the exercise of the human will. In this sense Shaw was an intellectually sophisticated Horatio Alger, carried over from the nineteenth century and raised to the nth power. It was not only possible to rise up from bootblack to riches (Shaw himself had virtually done that); it was also possible to rise up from the state of being "a feverish selfish little clod of ailments and grievances complaining because the world will not devote itself to making you happy" (II, 523) to the state of glorious superman.

For Huxley and his followers, as we have noted, eugenics was to provide salvation for the world and assurance for its future. We are to control the world's population, make it more intelligent and more perceptive, by wiser and more careful choices in mating. Indirectly this will solve the world's other problems too if we are lucky enough to survive for the number of generations it will take to produce sufficiently superior people. The human race, Huxley reminds us, is not only exceedingly young (in terms of biological time), but also exceedingly imperfect; and if we are to survive and progress into the next higher phase of hominization (to use Teilhard's term) we have to improve our progeny and do so in a hurry.

Sir Julian does not see eugenics in the trite science-fiction context, with an all-powerful council determining how many farmers and how many engineers or economists shall be born this year. He sees human variability as the primary quality to protect and foster, and he is willing to trust to education and vigorous propaganda, backed by government help and approval, to provide the enlightenment needed to bring about results. In this he is encouraged by the progress of birth control during his own lifetime, recalling that so recently as 1917 Margaret Sanger was jailed for disseminating birth control information. Eugenics should move in two directions: a negative one which would aim at eliminating heritable diseases and mental defects; and a positive direction to promote the qualities needed for high civilization — intelligence, imagination, empathy, cooperation, a sense of discipline and duty. (These are his value-choices.) Both the positive and negative results of eugenics can be achieved by the practice of birth control, artificial insemination, and artificial parenthood, with the use of deep-frozen sperm and (eventually) ova.[1]

Huxley was continuously careful to point out, in dealing with our future, that evolution had ceased to be a purely biological matter, but had become "psychosocial." That is to say, the genetic variations that now count as improvements are those which help humans to survive, adapt, and propagate *in the environment which they have made for themselves*. It is the *phenotype*, in biological terms, which is significant, rather than the *genotype*. But though he charts an evolutionary path into the future, he does not construct or fantasize a future civilization. (One must turn to his brother, Aldous, for that.) He recognizes very clearly the social and economic problems that lie in the way, but there is an undeniable naiveté in his political outlook, and in his philosophical assumption that the attributes *he* has selected for human improvement will remain the obvious and universal choices of the rest of the earth's inhabitants.

On the subject of eugenics as a basis for the future of humanity, Teilhard does not press the point as far or in as great detail, possibly because he does not want to exacerbate his already difficult relationship with his superiors in the church. He differs from Huxley, of course, in his insistence that the practice of eugenics is all part of "a vast, *directed* movement, bound up with the very structure of the Cosmogenesis."[2] But though Teilhard is not as specific as Huxley in eugenic procedure, he is considerably more so in his vision of what these genetic changes must achieve. They must achieve, first of all, "social unification" — what he called elsewhere "planetization." Humanity must move toward becoming a single organism. In pursuit of this we must regard our technology as an extension of ourselves, just as Marshall McLuhan was to advise a couple of decades later. And such outer changes presuppose inner changes as well. Man's mind must be freed from its biases and set to its true tasks of understanding itself and the universe. There must be a simultaneous advance, in other words, of Society, the Machine, and Thought.

With these aims for humanity, it cannot be denied that Teilhard looked on the rise of the dictators, as Shaw and Huxley also did, with some hope. By 1948, however, he had to admit that "the individual, outwardly bound to his fellows by coercion and solely in terms of function, deteriorates and retrogresses." We must wait for more fundamental changes within the human psyche. We must wait, in short, for the Christian concept of love to bring people together, "using the word 'love' in its widest and most real sense of 'mutal internal affinity.'"[3] Teilhard's conclusion to his essay on "The Directions and Conditions of the Future" clearly illustrates both why he continued to consider himself a faithful, if advanced, Christian; and why his superiors in the church considered him theologically unsound and unsafe:

> For a Christian, provided his Christology accepts the fact that the collective consummation of earthly Mankind is not a meaningless and still less a hostile event but a pre-condition of the final 'parousiac'[4] establishment of the Kingdom of God — for such a Christian the eventual biological success of Man on Earth is not merely a probability but a certainty: since Christ (and in Him virtually the World) is already risen. But this certainty, born as it is of a 'supernatural' act of faith, is of its nature supra-phenomenal: which means, in one sense, that it leaves all the anxieties attendant upon the human condition, on their own level, still in the heart of the believer.[5]

Shaw's view of the future was not derived from direct scientific observation. Although, on occasion, he called himself a "metabiologist," such a designation was pretentious. He was in no sense a scientist. He admitted his own deficiency in mathematics, and I know of no instance where he performed a laboratory experiment, participated in an excavation, or in any way struggled with primary sources of information. He was, however, able to absorb and digest vast stores of information, to bridge highly disparate disciplines, and to create original hypotheses out of other people's ideas. Consequently, though he does not write as a scientist, he by no means writes in ignorance. He is more convincing when he refers to himself, as he frequently does, as a mystic. But he is not the sort of mystic who draws his ideas purely from visions. He is an *informed* mystic.

Shaw's vision of the future, while not wholly in conflict with either Huxley's or Teilhard's, presupposes a clearer post-human stage than either of them pictures. His view is less sanguine. He does not believe that the human animal as presently constituted is competent to advance civilization on the earth, or perhaps even to survive. Some radical change must take place in the evolutionary pattern, and it must happen even more quickly than could be brought about by eugenics. (He did not know, of course, about the possibility of altering the genetic information in the gene itself; and it is difficult to assess what he would have thought of this.) He wanted people to will a change — a quantum leap in the evolutionary process. The change would have to be in mentality or consciousness, and Shaw linked this change with longevity. He reasoned that people simply do not live long enough or keep their vitality long enough to attain the experience and the wisdom needed for taking charge of their own evolutionary future. Longevity, therefore, is a dramatic device to symbolize the evolutionary breakthrough into the new consciousness. It is the physical attribute that must somehow accompany the mental and spiritual change.

In Part II of *Back to Methuselah* ("The Gospel of the Brothers Barnabas") which is set in the then-present time of the early 1920s, Shaw presents his gospel of long life as a solution to the world's problems through a pair of brothers, who are, significantly, a biologist and an ex-clergyman. The play does follow Bergsonian philosophy in that the idea has to be *thought* before it can happen. "Originally," Bergson says, "we think only in order to act."

> Our intellect has been cast in the mold of action. Speculation is a luxury while action is a necessity. Now, in order to act, we begin by proposing an end; we make a plan, then we go on to the detail of the mechanism which will bring it to pass. This latter operation is possible only if we know what we can reckon on. We must therefore have managed to extract resemblances from nature, which enable us to anticipate the future. Thus we must, *consciously or unconsciously,* have made use of the law of causality.[6]

This could be the rationale for "The Thing Happens" — the title of Part III — just two hundred and fifty years after Part II, when a few scattered individuals who have been touched by Barnabas' ideas find themselves living for three centuries, with their bodies aging only as if they were living a normal lifetime. Shaw is seriously entertaining the notion, then, that there are those alive in the 1920s who are already marked as members of a select longer-lived race of humans.

Interestingly the Barnabas brothers themselves are not among the elect. Instead, the local clergyman of the previous play, who was only mildly interested in talking with the Barnabas brothers, and the parlor maid who had peeked into the Barnabas book, find themselves still alive and healthy in the year 2170. Each thinks himself or herself unique, and, embarrassed at the inability to grow old, has invented false deaths and new identities to avoid the countless implausible explanations that the situation would require with constantly aging contemporaries. Eventually, however, they meet and discover each other's secret. From that moment it is obvious that they have a destiny to fulfill. They must mate and genetically pass along the gift of longevity to their children.

Now if we regard the Shaw characters' sudden attainment of longevity as actual, we are faced with two conflicts in evolutionary theory. One concerns the possibility of a quantum jump, and the other concerns the motivating force of the human will. On either of these Shaw could expect no support from the neo-Darwinians. Huxley, it is true, does question Darwin's "Natura non facit saltum," but his acceptance of a "leap" would not encompass the sudden emergence of three-century-old humans in our midst.

> . . . whatever theory of variation we may hold — the old idea of small continuous variations; or that of large mutations big enough to produce new species at one jump; or the most probable theory of numerous small mutations — they one and all must grant that the largest variation occurring at one time in a living species is infinitesimal in comparison with the secular changes of evolution.[7]

This latter point is supported by more recent evidence. Stephen Jay Gould, in making a case for "speciation" — the splitting of one lineage from a parental stock — maintains that this process can be "*very rapid* by evolutionary standards," but goes on to say that he is talking about "hundreds or thousands of years (a geological microsecond)."[8]

But if the leap is symbolic and does not involve actual physical changes in the organism, Huxley can give some small philosophical support:

> We are now on the threshold of some . . . critical revolution of thought in which human ideology is destined to be radically reorganized, and our old patterns of ideas and beliefs will be superseded by a new dominant idea-system.[9]

However, Huxley's "critical revolution of thought" would be brought about by the practice of eugenics, by education, and by government action, not by some sudden shift to a longer life span.

The evidence is, however, that Shaw *does* regard the extension of human life as something more than a dramatic symbol:

> . . . the very vulgar proposition that you cannot change human nature, and therefore cannot make the revolutionary political and economic changes which are now known to be necessary to save our civilization from perishing like all previous recorded ones, is valid only on the assumption that you cannot change the duration of human life. If you can change that, then you can change political conduct through the whole range which lies between the plague-stricken city's policy of "Let us eat and drink; for tomorrow we die" and the long-sighted and profound policies of the earthly paradises of More, Morris, Wells, and all the other Utopians. [V, 632][10]

Teilhard does not specifically link future evolutionary development with longevity, but he does not rule out the possibility that there were in the past and may be in the future the kind of quantum leaps Shaw has dramatized. The emergence of life on the earth, for example, may have represented such a leap: ". . . we must postulate at this particular moment of terrestrial evolution a coming to maturity, a threshold, a crisis of the first magnitude, the beginning of a new order."[11] More pertinent, however, is the growth in human beings of *reflective consciousness.* I am not sure I understand the full meaning of this term which Teilhard uses quite frequently, but it refers to an outgrowth of the noosphere which is uniquely human, and has to do with the ability of consciousness "to fold back on itself," or to dwell upon itself. This makes it possible for man to establish what Shaw (along with Nietzsche and other nineteenth-century philosophers) called a *will,* though Teilhard himself steers clear of using that term. And this uniquely human power that man attains through the gift of reflective consciousness might conceivably result in unprecedented changes in himself — even biological changes. Teilhard's literary style is sometimes difficult (as you have no doubt already noticed), but I must nevertheless quote him here at some length. He has been remarking that the gradual growth of consciousness in animals does not seem to have increased the speed of their evolution, but that the appearance of man marked a radical change:

> For Man, by the act of 'noospherically' concentrating himself upon himself, not only becomes reflectively aware of the ontological current on which he is borne, but also gains control of certain of the springs of energy which dictate this advance: above all, collective springs, in so far as he consciously realises the value, biological efficiency and creative nature of social organization; but also individual springs in as much as, through the collective work of science, he feels himself to be on the verge of acquiring the power of physico-chemical control of the operations of heredity and morphogenesis in the depths of his own being. So we may say that since by a sort of chain reaction [,] consciousness, itself born of complexity, finds itself in a position to bring about 'artificially' a further increase of complexity in its material dwelling (thus engendering or liberating a further growth of reflective consciousness, and so on . . .) the terrestrial evolution of Life, following its main axis of hominisation, is not only completely altering the scale of its creations but is also entering an 'explosive' phase of an entirely new kind.[12]

He goes on to say that as evolution moves into the area of reflective consciousness, it becomes less and less "Darwinian" and more and more "Lamarckian." It ceases being passive and "*becomes active in the pursuit of its purpose.*"[13]

These last words of Teilhard's might well have appeared in the Preface to *Man and Superman.* Indeed they are a condensation of the thought that Shaw expressed through his Brothers Barnabas:

> The notion that Nature does not proceed by jumps is only one of the budget of plausible lies that we call classical education. Nature always proceeds by

> jumps. She may spend twenty thousand years making up her mind to jump; but when she makes it up at last, the jump is big enough to take us into a new age.... Spread that knowledge and that conviction; and as surely as the sun will rise tomorrow, the thing will happen.... Do not mistake mere idle fancies for the tremendous miracle-working force of Will nerved to creation by a conviction of Necessity. I tell you men capable of such willing, and realizing its necessity, will do it reluctantly, under inner compulsion, as all great efforts are made. They will hide what they are doing from themselves: they will take care not to know what they are doing. They will live three hundred years, not because they would like to, but because the soul deep down in them will know that they must, if the world is to be saved. [V, 429–33]

In the meantime the question of the speed of the evolutionary process and the possibility of evolutionary breakthroughs is far from settled. Arthur Koestler has assembled an impressive catalog of biological phenomena that cannot be accounted for by orthodox Darwinian reasoning.[14] And to these must now be added the discovery of a group of Italian scientists headed by Mario Coluzzi, who observed chromosonal changes in certain breeds of mosquitos, suggesting a kind of evolution that is rapid enough to be watched through a microscope.[15] We must remember constantly that the proportion of time during which we have had our own species under recorded observation, compared to the time the species has been in existence, is miniscule, and all pronouncements on the detailed nature of our early development are, at best, carefully considered hypotheses.

Except for its speed and its motive, the Shavian version of evolution does not differ radically from orthodox Darwinism. If the next step of evolution *were* to develop longer-lived people (which may be happening at a slower pace), the development would not likely be consistent. Only a small portion of the population would be affected. Huxley reminds us that the simplest forms have always survived side by side with the more complex, the less specialized with the more specialized, the lower types alongside the higher.[16] So the presence of longer-lived and shorter-lived in the same culture would not be exceptional. Furthermore, Teilhard points out that the more advanced types in such a population recognize and are attracted to one another,[17] just as the two longlivers are in "The Thing Happens."

Once the longlivers become conscious of their mission in *Back to Methuselah*, progress is even more rapid. By the year A.D. 3000 — in fewer than thirty generations — an entire new civilization of longlivers has isolated itself (in Ireland!) from the rest of humanity, which must go plodding along with its allotted three score and ten. In the accumulated wisdom of a community in which everyone can expect to live for three centuries there are few recognizable human concerns. As Shaw points out, it is not the mere fact of longevity that creates wisdom; it is the *expectation* of longevity. If we could regard the age of sixty, say, as adolescence, and know that, barring accident, we had at least a couple of centuries more of productive life ahead of us, our entire set of values would change. The Barnabases, as Shaw put it later,

> had come to the conclusion that the duration of human life must be extended to three hundred years, not in the least as all the stupid people thought because people would profit by a longer experience, but because it was not worth their while to make any serious attempt to better the world or their own condition when they had only thirty or forty years of full maturity to enjoy before they doddered away into decay and death. [V, 632]

We really see very little of the advanced civilization in Part IV of *Back to Methuselah*, partly because the action of the play necessarily keeps us on the fringe of it; and partly "because, being a shortliver, I could not conceive it."[18] The land does not appear to be over-populated. The dress and the architecture are classically simple. The inhabitants have developed a kind of communication by means of intoning on a single pitch — a kind of walkie-talkie. But Shaw is not at his best in imagining science-fiction technology. Writing seven years before Lindbergh's first transatlantic flight, Shaw has the Elderly Gentleman of A.D. 3000 talking of crossing the ocean in an afternoon and of instant visual communication as though these were recent developments; and he speaks of a universal abundance of leisure as though it were even then still in the future. He even speaks of having coal in the cellar, and there is not a hint of anything like nuclear energy or space travel. What Teilhard called the "ultra-technified" future is not apparent here. Shaw imagines, too, that the world after another millenium is still dominated by the British Commonwealth (with its capital removed to Baghdad) and that it faces a rival empire of "Turania." We may be unimpressed at such topical details of future-guessing, but the account by one of the longlivers of "the end of pseudo-Christian civilization" by a combination of bombing cities and using "a gas that spared no living soul" is more soberingly convincing.

The longlivers of Part IV are really a new species — competent, self-assured, with inbred cooperative instincts. Having found themselves incompatible with our ordinary posterity, they have taken over the islands of Ireland and its eastern neighbor, leaving the rest of the world to go on very little changed from what it was in 1920. The longlivers have set up a rational social organization divided into "primaries," "secondaries," and "tertiaries," according to whether they are living in their first, second, or third century, and have apparently solved all social, economic, and political problems to the point where they can devote themselves to living on a much more exalted plane intellectually and esthetically. They have left, as we shall see, only one "political" problem of real concern.

Almost all contact between the two worlds has been eliminated and what remains is carefully monitored. Part IV is titled "The Tragedy of an Elderly Gentleman" and has to do with an authorized visit of a group of shortlivers — a prime minister and his party — into the land of the now legendary supermen. The visitors wish to consult the "Oracle" about the advisability of calling a new election. The longlivers consent to play such games with them because humoring them is the only way they know of dealing with them.

The Elderly Gentleman of the play, the Prime Minister's father-in-law, becomes separated from the party and manages, under a succession of three different "caretakers," to learn more of the longlivers than is intended, and more than is good for him. The longlivers assume that all the old destructive emotions of hate and malice have been bred out of them, but Zoo, a mere child of fifty, in the process of dealing with the Elderly Gentleman, finds these unwanted emotions still latent in her. The Gentleman, frustrated at his inability to comprehend the values of the longlivers, loses his temper:

> THE ELDERLY GENTLEMAN [*thundering*] Silence! Do you hear! Hold your tongue.
> ZOO. Something very disagreeable is happening to me. I feel hot all over. I have a horrible impulse to injure you. What have you done to me?
> ....
> ... It is not merely that you threw words at me as if they were stones, meaning to hurt me. It was the instinct to kill that you roused in me. I did not know it was in my nature: never before has it wakened and sprung out at me, warning me to kill or be killed. I must now reconsider my whole political position. [V, 523–24]

Zoo, at fifty, is a primary. The secondaries are even more aloof, barely tolerating the visiting shortlivers. In a sub-plot of Act II, a special guest of the party, the Emperor of Turania, Cain Adamson Charles Napoleon, a man of great charisma in the shortlived part of the world, is humiliated and reduced to a quivering mass of fear at the full gaze of a secondary. We never meet a tertiary.

These are not quite the supermen we expected from the author of *Man and Superman*. John Tanner (Don Juan), Shaw's proto-superman, was certainly more brilliant, more entertaining, more fascinating. But Tanner's mission was to arouse and shock bourgeois society of ante-bellum Edwardian England, and not to be part of a productive super-society. Besides, Tanner was not really the superman any more than Ann Whitefield was the superwoman. They were, perhaps, marked by the Life Force to be early progenitors of the new race, but they were not destined to be a part of it.

Teilhard, too, foresees a definite "ultra-human" stage of evolution.

> ... we see Mankind extending within the cone of Time beyond the individual; it coils in collectively upon itself above our heads, in the direction of some sort of higher Mankind.[19]

Teilhard does not see his ultra-human society as being isolated, but as gradually spreading on a global scale, with a "nucleus of ultra-consistence" fusing the human multitude together. His vision is of "a higher state of consciousness diffused through the ultra-technified, ultra-socialised, ultra-cerebralised layers of the human mass."[20] Since he is not a dramatist, he is not obliged to materialize his vision upon the stage, but one can readily see that his dramatization would have to be substantially different from Shaw's. We would have a scene where the ultra-humans were woven inextricably into the social fabric that included the ordinary humans, but where the ultra-humans were making their superiority apparent through their advanced technology, sense of organization, and mastery of abstract thought. And as a man trained in science and the pursuit of truth, he foresees

> a world in which, not only for the restricted band of paid research-workers, but also for the man in the street, the day's ideal would be the wresting of another secret or another force from corpuscles, stars, or organized matter; a world in which, as happens already, one gives one's life to be and to know, rather than to possess.[21]

But he did not anticipate the eventual and inevitable conflict between the humans and the ultra-humans, as Shaw did.

Such a conflict is the central theme of "The Tragedy of an Elderly Gentleman." The Elderly Gentleman is, of course, not a man of the future at all. He is simply one of us, living socially and politically in the pattern of the twentieth century, with a twentieth-century set of values. At home, among his own kind, he is an educated, well-traveled man, a liberal thinker, and a political leader. Here, among the longlivers, his self-assurance completely collapses. He is made to seem stupid only because he is in the presence of hopelessly superior beings. When he is not interrupted by Zoo or Zozim he sounds very much like Tarleton, the well-read and self-possessed underwear manufacturer of *Misalliance*. But he has neglected a Secondary's warning that "There is a deadly disease called discouragement, against which shortlived people have to take very strict precautions. Intercourse with us puts too great a strain on them" (V, 493). He is shaken to find that these people have abandoned all concepts of, and therefore have no words for, decency, pauper, trespassing, private property, landlord, sneer, moral sense, insult, marriage, liberty. They have no need for sleep. They have no need for metaphors or images. Romanticism or any form of coloration of the truth is incomprehensible to them. Extended exposure is bound to make the shortliver realize that "we are worms before these fearful people: mere worms" (V, 544).

Part of the aura of frustration which pervades the attempt to communicate between the two worlds is a result of the complacent manner in which the Elderly Gentleman misreads history. History to him is a pleasant blur to which he frequently refers in order to validate his own opinions. By the year 3000 even the names have become blurred. The "father of history" turns out to be Thucyderodotus Macollybuckle who is credited with *The Perfect City of God*. The ancient religious reformer is Jonhobsnoxious. And the accounts of events, both of the pre- and post-twentieth century, are no sharper. Zoo is prompted to ask, "Why do you shortlivers persist in making up silly stories about the world and trying to act as if they were true?" (V, 511).

It is our own world, of course, that is false and unreal and hopeless. The future lies not with the likes of us, but, if at all, with a new race of creatures who may emerge from us and may resemble us in superficial ways, but will be unimaginably better able to realize themselves and handle their own affairs. We are the prisoners of our limited years and we cannot pass on "the ever-burning torch" from generation to generation as we fondly pretend, but must begin again and again with the same lack of wisdom. When the Elderly Gentleman declares that "human nature is human nature, long-lived or shortlived, and always will be" (V, 520), he has not yet grasped the meaning of long life-expectancy. And since we in the audience know that we are a part of the Elderly Gentleman's world, despite the thousand years projected between us, the effect of the play is to show us how our world looks through superior and detached eyes.

Some of the effect is humorous. A number of the scenes can be played for laughs. But the humor is indeed dark. The Prime Minister and his party, including the Elderly Gentleman, are ridiculous in their fright, subservience, and pettiness in the presence of the longlivers. There is some vaudeville business with a gun, when the "Oracle" shoots Napoleon — and misses — in an obvious bit of comic relief. Occasionally there are the expected Shavian witticisms. When Zozim asks the Elderly Gentleman what a church is: "You must excuse me; but if I attempted to explain you would only ask me what a bishop is; and that is a question no mortal man can answer" (V, 547). And there are a number of humorous topical references. But all the chuckles are technically induced. The spirit of fun is a mere echo from the twentieth century.

Unhappily for him, the Elderly Gentleman is far more perceptive than the other members of the party. To humor the shortlivers on these occasions the longlivers stage an impressive show in which a Secondary woman plays the part of the Oracle. It soon becomes apparent that the Oracle has no interest in the petty political questions that the Prime Minister has brought her, and her only response is, "Go home, poor fool" (V, 560). The party is then faced with the problem of how to report the results of their visit to the electorate. There is nothing to do but lie, as, they now realize, all their predecessors must have done. But the Elderly Gentleman finds he can no longer participate in such a falsehood. "I cannot live among people to whom nothing is real. I have become incapable of it through my stay here." He is therefore faced with the choice of staying in the land of the longlivers and dying of discouragement, or returning home to die "of disgust and despair." He takes the "nobler risk," and the Secondary who played the Oracle mercifully looks him full in the face, and so puts an end to his life (V, 562–63).

The one political problem which the longlivers had not yet solved was the question of what to do with the shortlivers. On this issue the advanced civilization is split into two parties: "The Conservatives hold that we should stay as we are, confined to these islands, a race apart." The Colonizers wish to increase in numbers, and spread throughout the world, superseding and exterminating the shortlivers. As a result of the feelings aroused in her by the Elderly Gentleman, Zoo changes from the Conservatives to the Colonizers. She concludes, "You only encourage the sin of pride in us, and keep us looking down at you instead of up to something higher than ourselves. . . . Our true destiny is not to advise and govern you, but to supplant and supersede you" (V, 525–27). This is the most debilitating kind of "future shock" — one that goes beyond Alvin Toffler's concept — the realization that one's life, one's life style, one's entire civilization may count for nothing in the larger scheme of things.

There is no indication in the course of the play what political decision was reached concerning the shortlivers. But in Part V of *Back to Methuselah*, after 27,000 more years, and after who knows how many civilizations have come and gone, there is a whole new race of beings. They hardly resemble the longlivers of the play before, and there is no indication that we shortlivers have ever existed.

In dealing with the phenomenon of the extinction of races, Charles Darwin noted that primitive races do not survive long in the presence of more civilized societies. And it is not merely because the Europeans bring new diseases and bad habits with which the primitives cannot cope. He cites an eminent anthropologist as laying "great stress on the apparently trifling cause that the natives become 'bewildered and dull by the new life around them; they lose the motives for exertion, and get no new ones in their place.'"[22] Would it not be proper to suggest, as Shaw does, that they die of discouragement?

12

"As Far As Thought Can Reach"

Sir Julian Huxley, in a critical article concerning Arnold Toynbee's use of time-scales,[1] wisely cautions us about confusing sidereal time, geological time, or biological time with historical time. In terms of sidereal time, the solar system may be very young; in terms of geological time, the history of life may seem very short; in terms of biological time the history of mankind seems but a moment, and measured against sidereal time we are practically non-existent. But Huxley sensibly reminds us that we exist in historical time, where the span of a generation does count for something, and that we had better know our place and curb our pretensions to something short of the infinite. Consequently Huxley's probes into the future are cautious, and end where scientific forecast turns into sheer speculation.

But Teilhard de Chardin's organic interpretation of the Phenomenon of Man cannot stop short of "a reasonable forecast as to our future destiny, and the fate which is reserved for us at the end of Time."[2] He shares that compulsion with Shaw, who entitled the final play in his pentateuch, "As Far As Thought Can Reach." Inevitably there will be an end to civilization, to the human race, to the earth, and to the earth's supporting star, the sun. For Huxley these events are of primary interest in our understanding of the universe, but have little to do with humanity within any foreseeable historical time-frame. But for both Teilhard and Shaw the fact that these ultimate events are unimaginably distant does not invalidate the basic philosophical questions that they pose for us: Is there an end to human consciousness? If so, of what use is it to try to read a meaning into life in the interim? If not, what can be the nature of human survival beyond the physical end of the earth, and are we in any sense a meaningful part of that survival? Shaw shocked his 1910 correspondent, Leo Tolstoy, by asking, "Suppose the world were only one of God's jokes, would you work any the less to make it a good joke instead of a bad one?"[3] The question would not have occurred to Père Teilhard in quite that form, but he nevertheless considered the two alternatives:

> a. That the planetary collectivisation which confronts us is a crude phenomenon of mechanisation or senescence which will end by dehumanising us;
> b. That on the contrary it is a mark and an effect of biological super-arrangement destined to ultra-personalise us.[4]

Obviously he chose the latter view.

Since both men believe in a Lamarckian Life Force that gives direction to the evolutionary process, neither could be satisfied with answers that would be limited to our own tiny corner of the universe. For both, the extension of consciousness as far as thought can reach took them to a state beyond the need for "a local habitation and a name." Each projected a bodiless state of consciousness, for Teilhard an "Omega Point," for Shaw a "vortex of pure thought." Both assume that the mystical stuff of human consciousness is not limited to the surface of the earth or by the time span of human history, but is part of the purpose of an evolving universe and will continue beyond any changes in matter. In this sense (and only in this sense) Shaw can believe in "the life everlasting." He feels no need, as Teilhard does, to reconcile this concept with any Christian theology of the afterlife.

Thirty years after he wrote this play, Shaw, at the age of ninety-three, thought perhaps he should have called it "As Far as *My* Thought Could Reach" (VII, 413). In placing the time at A.D. 30,000, he doubtless felt that he was extending his vision beyond any imaginable limits. Today he might wish to revise his figures, since the earth-launched Voyager 2 is scheduled to pass by a particular star in the Little Dipper in the year A.D. 46,230 and into the constellation of Sagittarius in A.D. 149,380! But the year is really immaterial. The play takes place at the edge of human imagination. Since we can have no visual image of what a society at such a time would be like, Shaw again relies on a neutral classic framework. His picture is pastoral. The dress and architecture resemble "that of the fourth century B.C." We are in a sunlit glade with "a dainty little classic temple." There are flute players and graceful dancers. The shortlived people of our own world have completely disappeared and are remembered only as distant mythology, and the longlivers have extended their years almost to immortality. The only struggle now is between the youth — who hatch from external eggs, full-grown and with the power of speech, and remain somewhat adolescent until they are three or four — and the elders or "Ancients." In these first three or four years the youths repeat in rapid sequence the entire history of the race, passing through periods of selfishness and play and love-making, until they are ready for the serious business of life, which is thinking. This is Shaw's extrapolation of the theory that "ontogeny recapitulates phylogeny," and Teilhard might have gone along with this fanciful treatment of it.

But we have already noted that Teilhard's view of the future was neither pastoral nor classic. The future projected to the edge of his imagination would be one of "the optimum use of the powers set free by mechanization. . . . geo-economy, geo-politics, geo-demography; the organization of research developing into a reasoned organization of the earth."[5] Shaw's ultimate society tended to resemble the utopia of his old friend, William Morris, in *News from Nowhere*. Teilhard's would have been more like the projections of Buckminster Fuller.

But neither man's vision is of a utopia — a final stasis at which society has arrived. For both, the universe is dynamic, not static, and however distant the view they take, it does not stop there. They both dare to look beyond the end of human history, and when they do they are in harmony with one another. For both recognize an incorporeal element as the essence of humanness — reflective thought, the spirit, the soul — and it is this that will go on seeking interaction and fulfillment in the expanding universe.

It must be remembered that for Teilhard the irreversible process of evolution must go on *uniting* everything into ever more complex forms. His "Omega point" is where all consciousness must eventually converge. This convergence, he assures us, is not a process of abandoning our individual personalities, but one of developing an ultra-personality, which is like entering a kind of Godhead. The unifying forces in evolution he equates with the Christian concept of love; and he sees the Omega point as one of the attributes of Christ. Humanity will move to the Ultra-Human, as we have seen, and the Ultra-Human will accede "to some sort of Trans-Human at the heart of things."[6] But this trans-human Omega point need not be thought of as some distant point in time. It exists now, just as the beginning, the Alpha point, exists now. The mystics have always been in touch with the Omega point. As the quality of mysticism increases with the planetization of humanity, will it not become conceivable, Teilhard asks,

> that Mankind, at the end of its totalisation, its folding-in upon itself, may reach a critical level of maturity where, leaving Earth and stars to lapse slowly back into the dwindling mass of primordial energy, it will detach itself from this planet and join the one true, irreversible essence of things, the Omega point? A phenomenon perhaps outwardly akin to death; but in reality a simple metamorphosis and arrival at the supreme synthesis. An escape from the planet, not in space or outwardly, but spiritually and inwardly, such as the hypercentration of cosmic matter upon itself allows.[7]

Teilhard's Omega point has obvious Biblical eschatological overtones —

"I am Alpha and Omega, the beginning and the ending, saith the Lord, which is, and which was, and which is to come, the Almighty."[8]

Shaw's phrase for the ultimate convergence of consciousness is the more secular "vortex of pure thought." Perhaps such a concept could be better expressed in science-fiction film images, somewhat in the manner of Stanley Kubrick's *2001*. To dramatize it with actors on a stage, Shaw made the young people of his far-distant society not too different from ourselves — recognizable in their preoccupation with sport and dance and love-making and art. They will eventually mature into "Ancients," but until they pass a certain point in their lives they do not understand the Ancients any more than we should. The Ancients have lost all interest in corporeal reality. They have experimented with changing their bodies, growing extra heads and eyes and arms (none of which we see on the stage), but now, after eight or nine centuries, literally as old as the Biblical Methuselah, they are tired of their bodies and wait for the eventual accident — the drowning, the fall, the lightning — which will release them. Two of them condescend (for the drama's sake) to converse for a few minutes with younger people in various stages of maturity. Martellus is a youth who has been showing signs of growing up:

> MARTELLUS. The body always ends by being a bore. Nothing remains beautiful and interesting except thought, because the thought is the life. Which is just what this old gentleman and this old lady seem to think too.
> THE SHE-ANCIENT. Quite so.
> THE HE-ANCIENT. Precisely.
> THE NEWLY BORN [*to the He-Ancient*] But you can't be nothing. What do you want to be?
> THE HE-ANCIENT. A vortex.
> THE NEWLY BORN. A what?
> THE SHE-ANCIENT. A vortex. I began as a vortex: why should I not end as one? . . . None of us now believe that all this machinery of flesh and blood is necessary. It dies.
> THE HE-ANCIENT. It imprisons us on this petty planet and forbids us to range through the stars.
> ACIS. But even a vortex is a vortex in something. You cant have a whirlpool without water; and you cant have a vortex without gas, or molecules or atoms or ions or electrons or something, not nothing.
> THE HE-ANCIENT. No: the vortex is not the water nor the gas nor the atoms: it is a power over these things.
> THE SHE-ANCIENT. The body was the slave of the vortex; but the slave has become the master; and we must free ourselves from that tyranny. It is this stuff [*indicating her body*], this flesh and blood and bone and all the rest of it, that is intolerable. Even prehistoric man dreamed of what he called an astral body, and asked who would deliver him from the body of this death.

> ACIS [*evidently out of his depth*] I shouldnt think too much about it if I were you. You have to keep sane, you know.
>
> *The two Ancients look at one another; shrug their shoulders; and address themselves to their departure.* [V, 623]

Though they do not say so, Shaw's Ancients seem willing to surrender their individual personalities to something more cosmic. They appear to have discovered, with Teilhard, that the ultimate consciousness is the universal consciousness.

In one of his most eloquent pieces of writing, at the end of "As Far As Thought Can Reach," Shaw calls back the ghosts of the characters from "In the Beginning": Adam, Eve, Cain, and the Serpent, and finally "the one that came before the Serpent," Lilith, who did not appear in the first play, but who is Shaw's personification of the Life Force. "What do you make of it?" she asks Adam, who passes the question along. They do not know what to make of it, and they do not agree. Only the Serpent declares, "I am justified. For I chose wisdom and the knowledge of good and evil; and now there is no evil; and wisdom and good are one." But it is Lilith who knows what to make of it. For her, even the Ancients are infants, and she knows that in their reaching out they will eventually "become one with me and supersede me, and Lilith will be only a legend and a lay that has lost its meaning."

> Of Life only there is no end; and though of its million starry mansions many are empty and many still unbuilt, and though its vast domain is as yet unbearably desert, my seed shall one day fill it and master its matter to its uttermost confines. And for what may be beyond, the eyesight of Lilith is too short. It is enough that there is a beyond. [V, 628–31]

The eloquence of Lilith's peroration can scarcely be matched by the translated prose of Teilhard de Chardin, but the nature of his vision is the same:

> . . . of thought becoming number so as to conquer all habitable space, taking precedence over all other forms of life; of mind, in other words, deploying and convoluting the layers of the noosphere.[9]

13

In Summary

In their view of the nature of human evolution and human destiny, Teilhard and Shaw differ from Huxley and the Darwinians as mysticism must always differ from purely rationalist humanism. But Teilhard is both a mystic and a scientist, and his view of pre-human evolution corresponds quite closely to Huxley's. Shaw seems to think that acquired characteristics — habits — can, through an act of will, be inherited. On the basis of scientific observation Teilhard must side with the Darwinians and say that there is no evidence for this. He does not dwell on the power of the lower life-forms to will their own changes; and he would probably not differ from the later Darwinians in the time-frame required for the gradual evolution of species. Nevertheless, he always leaves room for the operation of a universal consciousness in the evolutionary process. The presence of mind and consciousness throughout the universe helps to direct the combination of atoms and molecules and living cells in fulfilling their divine purpose. Though the acquired habits may not of themselves be inherited, Teilhard maintains that the continuous joining together of elements to make more and more complex organisms does not happen purely by chance. Something like choice is exercised by the smallest particles of matter, or by energy before it becomes matter, even though this choice is not observable in the pre-human phases of evolution.

Consequently in basic philosophy — we can almost say theology — Teilhard is really much closer to Shaw than to the Darwinians. As Daniel Leary has put it, "... both men were exponents of a perennial religion that predates Christianity, that indeed has its roots in the archetypal imagination of the human race and that seems now to be forcefully pushing forward into the consciousness of modern man."[1] Huxley agrees with Shaw and Teilhard that in the human phase of evolution man must be in charge of his own progress — up to a point. What constitutes progress must, for Huxley, be determined on a purely rational system of ethics. This is not sufficient for either Teilhard or Shaw, who conceive of morality as being in harmony with the spiritual source that directs the universe.

All three accept the possibility of human failure, of the destruction of humanity before it can move forward to a higher level. Shaw, on this point, is the most alarmist. Teilhard cannot quite conceive that the will of God would be left undone. Neither can Shaw, really, but he can imagine the complete failure of the human race to do it. For Shaw, as a non-Christian, humanity occupies no privileged position in the universe. If it cannot fulfill the purposes of the Life Force, some other creature in some other time-span or on some other planet must be evolved to do it. For the Christian, Teilhard, man is the especial child of God, and his failure would be a cosmic tragedy of Titanic scope.

It is regrettable, though really not surprising, that Père Teilhard was unable to obtain the *imprimatur* for the publication of his most important works during his lifetime, or to lecture freely about his evolutionary ideas. His scientific thought was, and probably still is, in conflict with Catholic dogma. He nevertheless continually strove to accommodate his observations to Christian beliefs. The stress is frequently evident. In the end he had to call for a revision, not only of scientific thought, but of Christian belief as well.

On the other hand Shaw was not bound by any creedal restrictions. In addition to being a dramatist — a world-famous one by the time he wrote *Back to Methuselah* — he could lay claim to being a social and political scientist. He had read exhaustively in social and economic theory, had been one of the guiding spirits of the early Fabian Society, served as Municipal Councillor for the Borough of St. Pancras, and was about to undertake a book-length exposition on politics and economics (*The Intelligent Woman's Guide to Socialism and Capitalism*). He was also skilled in argumentation, both on the platform and in print. It was certain that whatever he said would be said effectively and would be widely read. But he was not a biologist. His ideas about evolution lack precision and authority. What Huxley said about Bergson, he might also have said of Shaw: that he was a good poet but a bad scientist.[2]

Nevertheless Shaw's view was much needed at the end of "the infidel half-century." And it is unfortunate that Teilhard de Chardin's *The Phenomenon of Man* could not have been published in 1938 when it was written. Members of the scientific community dismissed Shaw's evolutionary ideas lightly, first because he was a mere *littérateur,* and secondly because they assumed his entire thesis rested on the inheritance of acquired characteristics, an abandoned hypothesis. But Teilhard, as is evidenced by Huxley, could not be so easily dismissed. And Teilhard made it clear that, quite apart from the argument over acquired characteristics, a Life Force *could* have been operating throughout evolution. The support has come late, but it is none the less welcome.

Of course neither Teilhard nor Shaw nor anyone else can prove the existence of a Life Force. But Teilhard, with his sweeping teleological view of the universe, has made it more difficult than ever to regard the mechanistic serendipitous mutation of organisms as the only means by which Life keeps improving itself. The great quantum leaps into the future that Shaw conceived may or may not be possible, but at least we have the choice, supported by some evidence, to believe that our journey has not been random, but has a direction. If we add the evidence together, Teilhard queries, "(and rectify certain exaggerated views on the purely 'germinal' and passive nature of heredity), how is it we are not more sensitive to the presence of something greater than ourselves moving forward within us and in our midst?"[3] Shaw would be willing to join Teilhard in pursuing that question. Huxley and the later Darwinians would not.

Part III

Ambiguity and Anguish

14

The Web of Ambiguity

What would seem most likely to cut Bernard Shaw off from the modern mind is his apparent lack of ambiguity — that pervasive quality that has increasingly colored all art in the twentieth century. Any of our contemporaries who offer clear answers are at once suspect. We may accept a qualified sense of direction in terms of social protest from an embattled Russian or from our own assertive minorities, but this must be largely a negative assertion against repression or tyranny. A clear position on the positive virtues marks a late twentieth century writer as passé. Shaw's stubborn and unambiguous belief in the twin forces of progress — Socialism and the Life Force — would seem, therefore, to have banished him from the consciousness of our times except as a historic phenomenon.

Shaw (the Broadbent-Shaw, the Doyle-Shaw) was an orderly person, with regular and disciplined work habits and a true admiration for efficiency. His early dramaturgy reflects this fastidiousness. He later claimed that he never knew the outcome of a play when he started it, but the finished work always had a recognizable structure, even though the pattern was not always conventional. Borrowing from Ibsen, he noted that "Formerly you had in what was called a well-made play an exposition in the first act, a situation in the second, and an unraveling in the third. Now you have an exposition, situation, and discussion; and the discussion is the test of the playwright."[1] Last-act discussions had, in fact, become accepted and expected in Shaw plays — in *Candida*, in *John Bull's Other Island*, in *Major Barbara*, and other less obvious examples. In *Getting Married* and *Misalliance* the discussion runs as a thread concurrent with the action, uninterrupted by change of scene or time. *Fanny's First Play* is a play within a play. The bold third act of *Man and Superman*, "Don Juan in Hell," is a dramatic metaphor that has its own structural integrity. These departures from conventional play form were received as novelties or innovations at the time, but no one today would consider them either radical or formless. They would more likely be regarded as models of structure. What is less obvious is that Shaw's passion for order always concealed deep and painful uncertainties which finally overwhelmed the dramatist as he watched all Europe devastate itself in the Great War.

Shaw has a way of making his audience feel temporarily gratified at the conclusion of a play — a kind of sounding of a good tonic chord. But the satisfaction does not last. There are other notes in the harmonic composition — deceptive notes that linger in the inner ear and remain to challenge the original feeling of fulfillment. Though his more popular pre-war plays may seem at first glance to be as conventionally plotted as anything by Eugene Scribe, all of them lack the conventional "happy endings."

The projected marriages are inauspicious: Bluntschli's conquest of Raina in *Arms and the Man* probably comes as close to a romantically satisfying ending as Shaw ever got; but it must be remembered that the Bulgarian girl fell in love with a romantic illusion — a "chocolate-cream soldier" — and finds herself marrying an irresistable realist. At the end of *You Never Can Tell* the prospect of a happy union between Valentine and Gloria is fairly stated by the Waiter: "... it often turns out very comfortable, very enjoyable and happy indeed, sir — from time to time. *I* was never master in my own house, sir: my wife was like your young lady: she was of a commanding and masterful disposition ... " (I, 794). Judith, at the final curtain of *The Devil's Disciple*, is not united with the man with whom she fell in love when he sacrificed himself, but must remain married to the parson who disillusioned her. In *Man and Superman*, John Tanner and Ann Whitefield (like Valentine and Gloria) can only look forward to the continuous struggle for dominance. The union of Tom Broadbent with his Irish "heiress" (in *John Bull's Other Island*) is one of convenience for him and of sad necessity for Nora. What the home life of Barbara Undershaft and Adolphus Cusins will be challenges anyone's imagination. Jennifer Dubedat in *The Doctor's Dilemma* is a loving and faithful wife, but she is married to a faithless scoundrel, and on his death undertakes a dutiful but loveless remarriage. For Edith and Cecil in *Getting Married*, the holy rite of matrimony is finally performed in the empty church; but after the stormy prelude and the dismissal of the wedding party, it is possibly the most cold-blooded and rational union ever celebrated in the theatre. In *Misalliance* (the title might almost speak for all these unions) the spoiled Hypatia prevails upon her indulgent father to "buy the brute" for her, and so acquire Percival; while the Polish daredevil, Lina Szczepanowska, prepares to fly off with the terrified Bentley in a damaged 1909 airplane.[2]

The love-affairs are not consummated: Vivie Warren chooses independence and sends the ardent Frank packing. Candida, after an interlude of symbolic foreplay, sends the poet Marchbanks out into the night. Caesar leaves Cleopatra on the shore to wait for Anthony. Lady Cicely, after achieving the "conversion" of Captain Brassbound, "escapes" from his personal magnetism. Dr. Colenso Ridgeon is denied possession of Jennifer Dubedat though he has, in effect, committed murder for her. Lavinia, in *Androcles and the Lion,* chooses to remain faithful to her religion and rejects the handsome Roman captain — though she will allow him to come and debate with her sometimes (perhaps Shaw's notion of a happy ending!). Eliza does *not* come back to fetch Henry Higgins's slippers at the end of *Pygmalion,* as she was made to do in *My Fair Lady.*

The lures of power and riches prove deceptive: The wealth of Mrs. Warren and her partners cannot change Vivie's mind. The military might of Rome and the genius of Julius Caesar only partially and temporarily hold Egypt in check. It cannot even contain the meek tailor, Androcles, when he is in the company of his friend, the Lion. Brassbound's brigands are no match for the sophistication and moral force of Lady Cicely Waynflete. Even the powerful military-industrial complex controlled by Undershaft faces an uncertain future when the next generation of Undershafts (Barbara and Cusins) takes over.

All these references to plays written before World War I offer sufficient evidence that the web of ambiguity had already formed around Shaw's thinking. Even then the sense of "you never can tell" made him eligible for twentieth-century consideration, and set the endings of his plays apart from the pat denouements of Arthur Wing Pinero, Henry Arthur Jones, W. S. Gilbert, and, to a lesser extent, perhaps, from those of Oscar Wilde and John Galsworthy.

It is strange, then, that *Saint Joan,* coming after *Heartbreak House* and *Back to Methuselah,* is generally regarded as the clearest and most *un*ambiguous of Shaw's plays — the straight story of a religious martyr, confused only by the aberrant Shavian epilogue. It is, on the surface, Shaw's most acceptable dramatization of the Life Force at work. Leon Hugo says, for instance, that "Shaw's Joan is the Life Force brought to an extraordinary pitch of intelligibility and is yet [he adds] ultimately intelligible only in itself."[9] Whether or not one can believe in the literal saints who spoke to Joan, it is obvious that she was inspired to throw herself in the face of the most powerful establishments of the medieval era, the Church and the State, and though her body was burned, her spirit, like Shakespeare's Caesar's, conquered.

But all this is a surface view, an emotional reaction immediately following a stirring production, with Joan's words at the trial still ringing in our ears (for this is eminently one of Shaw's most actable plays). But on another level it soon becomes evident that this play, too, is shot through with the most painful post-war ambiguities, and that Joan's final cry of "How long, O Lord?" obscures an even more despairing "Quo Vadis?"

Shaw designated *Saint Joan* as a chronicle play. But the chronicle would be just as complete without Scene 4, the first third of Scene 6, and the Epilogue. Since, in playing time, these three segments comprise over an hour's worth of the total three and a half hours, the designation is certainly inadequate. Shaw, whatever his first intent, soon found himself involved in something more than chronicling the life of a fascinating French saint-heretic. It becomes evident from the first scene that Shaw is on the trail of a larger idea when Joan tries to explain to the stupid de Baudricourt, "We are all subject to the King of Heaven; and He gave us our countries and our languages and meant us to keep them." (VI, 93). And in the following scenes with the Dauphin she again presses home her mission to drive the English from the soil of France, not merely as farsighted political strategy, but as an inspired message coming directly from the throne of God. The brief Scene 3 at Orleans is strictly chronicle, as the wind miraculously shifts at her arrival so that the armed rafts may be brought upstream to join Dunois' forces in their assault on the bridge to relieve the siege.

And now Shaw resorts to a most unusual departure in dramatic structure. He breaks into the chronicled narrative at the high point of Joan's success with thrity-three minutes of argumentative talk, which forms structurally the hinge of the drama. The argument, carried on within a tent in the English camp, is between the French Bishop, Cauchon, and the English warrior-nobleman, the Earl of Warwick, with anti-intellectual comic relief from the Chaplain de Stogumber. The exchange is passionate, since it involves what is closest to the heart of each — the security of the Church, the security of the peerage, and the unobstructed rights of English imperialism. The last-named could not be taken very seriously as a philosophical position even in 1923, since its mouthpiece, de Stogumber, is merely, at this point in the play, a blatant jingoist. But the other two make their points with great force, establishing Joan as a militant Protestant a century before Luther, and a proto-nationalist more than a century before Henry VIII's break with Rome. In these matters we know, as the characters of the play do not, that Joan was on the side of history; and from here onward the play broadens into a struggle between the old world and the new. We see Medievalism straining desperately to hold back the emergence of the Renaissance and the world that grew from it.

When we see Joan again — in the cathedral after the coronation and in the great trial at Rouen — we see her in a new light. She carries within her — and in her simple way she seems to know that she carries within her — the seed of the future of the West. That is why the Epilogue is essential to the larger meaning of the play: because the seed did grow, and it was Medievalism which died. The Epilogue should not be played as a tricky Shavian afterthought. It is, in musical terms, a coda; in terms of nineteenth-century melodrama, an apotheosis. If we are to accept *Saint*

Joan as more than the chronicle of the life of an unusual girl, we must know that it was her spirit, and not the spirit of Cauchon or Warwick, that prevailed. "The world is saved neither by its priests nor its soldiers, but by God and his saints" (VI, 198). Not only must we learn from Dunois that the English have been driven from France, but we must hear Warwick affirm that "Your spirit conquered us, Madam" (VI, 202–03), and learn from the Executioner that as a Protestant, Nationalist, and Individualist she could not be killed: "Her heart would not burn. ... She is up and alive everywhere" (VI, 202). We must discover, too, that Cauchon, the Bishop who so piously pursued her in the zeal of his righteousness, was himself posthumously excommunicated, his body disinterred and flung into the sewer, while Joan, secure in the hearts of the common people, slowly made her way, bit by bit, back into the bosom of Mother Church, and in just under five hundred years achieved canonization.

Why is it, then, that the world is still not ready to receive her? Why do they all, one by one, withdraw from King Charles's dream — or Shaw's imagination — and leave her alone on the empty stage, wondering in death as she did in life why the world will not embrace her? One cannot really accept their foolish little excuses as they leave — the heretic is better dead, we are not good enough for you, political necessity, the Inquisition cannot be dispensed with, and so on. Is it not evident by now that she has been in the true path of history? Has she not been the forerunner of our own times?

Alas, yes! And it is here that the ambiguity of the contemporary mind interposes itself between Shaw's own view of history and the image of his most successful heroine. And on the way home after the play, or next day or next week, we realize that the web of ambiguity had been weaving itself throughout the play — or at least since Scene 4. The artist-philosopher Shaw had once again defeated the tractarian Shaw. True, Joan was the wave of the future, the modern saint triumphant. But that future is now the present, and we are haunted by doubt: Was the Renaissance the best that could have happened? Is what we call progress — through the Reformation and the growth of nation states and the Age of Revolutions and the growth of democracy and industrialization and the acquisition of overkill is all this to be preferred over what might have flowered from the Holy Roman Empire if it had achieved the unity of Europe which it sought? Can we be *sure* that Joan was the true embodiment of the Life Force?

It may be that Shaw was unable to face all the implications of these questions head-on, though Louis Crompton has noted that even in its form — "in its juxtaposition of comedy and saintliness" — it is "closer to the religious drama of the Middle Ages than to anything in Elizabethan or later theatrical literature."[4] Shaw, too, in his own way, was a modern saint, behind him a full lifetime committed to Fabian Marxism and the forward movement of the Life Force. He was never the hollow optimist that some surface-viewers of his works took him to be. He was, at most, a meliorist, and an increasingly sceptical one by the 1920s. He was forced to accept the possibility of human failure on a cosmic scale. Yet he never gave up hope for us. "Defeatism is the wretchedest of policies," was one of his last pronouncements.[5] We might take some catastrophic wrong turn tomorrow. Indeed, we are forever on the verge of doing so. "We live," he once wrote, "as in a villa on Vesuvius."[6] But that the fatal turn had been irrevocably taken — that the villa was already buried under five centuries of lava — that would be a fascinating situation for a Samuel Beckett story, but for Shaw it would have been too horrible to contemplate. It would mean that the struggle nought availeth, and that was an admission neither Joan nor Shaw could ever make.

But Shaw was not Joan. He was no simple country lass caught between the terrible forces of the Church and the Law, but one of the most wide-ranging and sophisticated minds to grow out of the turbulent nineteenth-century British civilization. He forcefully proclaimed in the Preface to the play that Joan's medieval persecutors were not villains. He leaves it merely implicit that the medieval view of the universe, for all its absurdity as viewed from our own times, may have been a useful one. In any case the defenders of Medievalism are given powerful arguments.

In one of the longest speeches Shaw ever assigned to an actor, the Inquisitor makes a reasoned case for the necessity of the Inquisition and for its essential mercy (if one can accept the premise that nothing is more horrible than heresy). But though it is a virtuoso display of argument, no one can expect that a modern audience will be convinced of the necessity for burning Joan. The arguments of the Bishop of Beauvais (Cauchon) are, however, of a different order. Though he carefully defers to the Inquisitor at the trial, he makes clear that he does not really fear the sensational excesses with which the Inquisitor has associated heresy — bands of wild women and men running naked, large-scale polygamy and incest. What is on Cauchon's mind is the "heresy that is spreading among men not weak in mind nor diseased in brain: Nay, the stronger the mind the more obstinate the heretic," the heresy that "sets up the private judgment of the single erring mortal against the considered wisdom and experience of the Church ... this arch heresy which the English Commander calls protestanism" (VI, 168).

And this brings us back to the play's hinge, to Scene 4, in the tent, where Cauchon's most telling argument for Medievalism is made. For Cauchon is not merely *against* heresy. He is *for* a stable and peaceful world, which he believes can only evolve around a religious center. There is no other such center than the Catholic Church, and those who stand in the way of this evolution, no matter how innocent and laudable they may seem, are in the way of that holy purpose and must be dispensed with. That is the real danger of Joan, and to her he links the other heretics, Mahomet, Hus, Wycliffe, and by implication Luther and Zwingli and all those who are yet to come.

> What will the world be like when The Church's accumulated wisdom and knowledge and experience, its councils of learned, venerable, pious men, are thrust into the kennel by every ignorant laborer or dairymaid whom the devil can puff up with the monstrous self-conceit of being directly inspired from Heaven? It will be a world of blood, of fury, of devastation, of each man striving for his own hand: in the end a world wrecked back into barbarism. For now you have only Mahomet and his dupes, and the Maid and her dupes; but what will it be when every girl thinks herself a Joan and every man a Mahomet? [VI, 135]

This presumption constitutes for Cauchon the unforgivable sin, "the sin against the Holy Ghost."

Though the Earl of Warwick's fears are, as he freely confesses, motivated by his concern for the power of the peerage, and he can lay claim to none of Cauchon's idealism, Joan's incipient Nationalism is no less threatening than her Protestantism; and the Bishop concurs:

> It is essentially anti-Catholic and anti-Christian; for the Catholic Church knows only one realm, and that is the realm of Christ's kingdom. Divide the kingdom into nations and you dethrone Christ. Dethrone Christ, and who will stand between our throats and the sword? The world will perish in a welter of war. [VI, 139–40]

Well, has the prediction been, so far, wholly inaccurate? And remember that it is made, and powerfully made, before the play is half over, and cannot be shaken out of our consciousness no matter how sympathetic we are to Joan or how much we rejoice in her posthumous victories and eventual sainthood. It does not lay to rest that awesome, unShavian possibility that though Joan's world has been, for the moment, victorious, it may be a world that has been doomed from the start; and that the Bishop of Beauvais, for all that we cannot accept his medieval theology, may have been more on the side of the Life Force than Joan was. After all, we cannot accept Joan's theology either. We are attracted to her because she is warmly human, a thorough activist, and against the Establishment — all attributes dear to our times. But they do not negate the historical perception of the old Victorian medievalist, Bishop Stubbs, as he weighed the Renaissance world against the medieval one:

> The sixteenth century, as a century of ideas, real, grand, and numerous, is not to be compared with the thirteenth century. The ideas are not so pure, not so living, nor so refined. The men are not so earnest, so single-hearted, so lovable by far. Much doubtless has been gained in strength of purpose, and much in material progress; but compare the one set of men with the

> other as men, and the ideas as ideas, and the advantage is wonderfully in favor of the semi-barbarous age above that of the Renaissance and the Reformation.[7]

Behind Shaw's rallying calls to follow him into the new land ("To Orleans!" [VI, 116]) one can sense the unspeakable dread that paradise may be lost and can never be regained ("My voices have deceived me. I have been mocked by devils" [VI, 179]).

With *Back to Methuselah* and *Saint Joan* Shaw had purged himself of the despair of *Heartbreak House*, but he had not purged himself of doubt. And doubt perceptibly colors the plays of his later years. None of them after *Saint Joan* achieved high critical acclaim; yet I agree with Katherine Haynes Gatch that "their reputation is in some measure the result of a failure to establish the critical bases on which these plays may be assessed as Shaw's peculiar contribution to English stage comedy in the second quarter of this century."[8] Their improvisational structures contain an ambiguous content puzzling enough in its existential complexity to belong to our own times. I believe that the painful sense of uncertainty in these later plays will communicate with today's audiences whenever the production is sensitive enough to allow it to do so.

In the 1930s Shaw found himself experimenting with what writers twenty years later were to call "serious farce." I have already referred to one of these later improvisations or extravaganzas in relating how Shaw's interest in the Quaker Conshies led him to create his *opéra bouffe* portrait of the Quaker founder in *"In Good King Charles's Golden Days." Good King Charles* was not written until 1939, but the prototype of the serious farce was written in 1931. Shaw called it *Too True To Be Good*, by which he meant, if you listen to the final speech of the play, that the characters are too lifelike to be believable on the stage: "There is something fantastic about them, something unreal and perverse, something profoundly unsatisfactory. They are too absurd to be believed in; yet they are not fictions: the newspapers are full of them . . ." (VI, 525). The story is a loosely structured farce: A rich and overprotected young lady is kept in bed with a microbe (the microbe is played by an actor) until she is convinced by her nurse and the nurse's burglar-boyfriend to allow herself to be kidnapped and to share the ransom which her wealthy mother will certainly provide. The adventure takes the three of them to a tropical British outpost where they pose as a countess (the nurse), and countess's brother, and a native servant (the rich young lady). The local colonel is completely taken in, but Private Meek — Shaw's comic portrait of Lawrence of Arabia — sees through everything. Meek, in fact, engineers phony battles with non-existent brigands to win the K.C.B. for the colonel. When in the midst of all this the rich young lady's mother arrives to find out what has happened to the ransom money and to reclaim her lost daughter, the succeeding events are as confusing and incredible as in the zaniest nineteenth-century French bedroom farce.

But there is a difference. Aubrey, the burglar, has been a wartime pilot who has dropped bombs on sleeping villages. He was brought up an atheist, but found he had a gift for preaching and has become an ordained clergyman. The nurse, Sweetie, is an earthy character with a high libido. If she has a man, her needs are satisfied. But the other characters are more complex, and Sweetie has difficulty, within the heavy intellectual atmosphere and with the limited number of males available, in arranging a suitable liaison. As Aubrey says,

> We all have — to put it as nicely as I can — our lower centers and our higher centers. . . . Since the war the lower centers have become vocal. And the effect is that of an earthquake. For they speak truths that have never been spoken before — truths that the makers of our domestic institutions have tried to ignore. And now that Sweetie goes shouting them all over the place, the institutions are rocking and splitting and sundering. They leave us no place to live, no certainties, no workable morality, no heaven, no hell, no commandments and no God. [VI, 477]

Sweetie abandons the burglar-clergyman for the more likely sergeant; but he, too, neglects his love-making to ponder on *The Pilgrim's Progress* and the Bible. Aubrey's father appears, an old-testament elder who has preached the doctrine of atheism and determinism. He finds that his son has become a thief, a clergyman, and a scoundrel, and that Einstein has destroyed the stable universe of Isaac Newton. The farce is not simply a confusion of going in and out of bedroom doors, but a confusion of who is in bed with whose ideas.

There can of course be no conclusion to such a plot except a hastily patched-up one. But the preacher-burglar continues to preach even after the play is over. "I am by nature and destiny a preacher," he declaims. "I am the new Ecclesiastes. But I have no Bible, no creed: the war has shot both out of my hands. . . . I must have affirmations to preach" (VI, 527). In a final stage direction Shaw adds that "*. . . fine words butter no parsnips. A few of the choicer spirits will know that the Pentacostal flame is always alight at the service of those strong enough to bear its terrible intensity*" (VI, 528).

One cannot avoid the temptation to contrast the pre-war Shaw, speaking, perhaps, in the pulpit of the City Temple, leaving his audiences with an inspired message of the Life Force, with Aubrey's rambling on into the darkness, searching for "affirmations to preach."

Most of these latter plays of the thirties and forties are, at base, political. They are concerned with the mystique of leadership, with what constitutes good government, and the necessity of achieving it before we destroy ourselves. The religious theme keeps recurring, and the presence of the

Life Force is never far below the surface; but it never erupts gloriously as it did at the end of *Major Barbara* or overcomes repression as it did in *Androcles and the Lion.*

One of his most improbable plays, *The Simpleton of the Unexpected Isles* (1934), is an allegorical fantasy set in tropical Oriental islands that rose mysteriously from the sea. The islands are an outpost of the British Empire, and the somewhat rambling story depends upon the culture shock that Westerners experience when they set foot in this strange environment. There is a long-standing eugenics experiment in progress here, which involves a complex arrangement of group-mating among Eastern and Western partners, and a confused English clergyman who is drawn into the experiment. The second act features an actual Judgment Day, during which the Angel of Judgment soars overhead, is shot at, but blithely lands onstage, shaking the lead bullets from his feathers. We are told that "The Day of Judgment is not the end of the world, but the end of its childhood and the beginning of its responsible maturity." Judgment means that "The lives which have no use, no meaning, no purpose, will fade out. You will have to justify your existence or perish. Only the elect shall survive" (VI, 825). The Angel, having left his message, flies off; and in due course word comes over the radio of people disappearing:

> At the Royal Institution Sir Ruthless Bonehead, Egregious Professor of Mechanistic Biology to the Rockefeller Foundation, drew a crowded audience to hear his address on "Whither have they gone?" He disappeared as he opened his mouth to speak. . . . The Times, in a leading article, points out that the extreme gravity of the situation lies in the fact that not only is it our most important people who are vanishing, but that it is the most unquestionably useful and popular professions that are most heavily attacked, the medical profession having disappeared almost en bloc, whilst the lawyers and clergy are comparatively immune. [VI, 834]

But those who have useful work to do in the world are not disturbed by the Day of Judgment, and the characters go back to their tasks, leaving us with the thought that "the future is to those who prefer surprise and wonder to security" (VI, 840).

As the second World War threatened and eventually became another terrible reality, the Life Force in Shaw's remaining plays took on a specific assignment, i.e., to raise up from the common mass of people leaders who would have the wisdom and the will to govern, and the magnetic personalities that would enable the people to accept them. Three of these predominantly political plays I have reserved for the next chapter. A few words need to be said about the appearance of "superior beings" in the other plays of Shaw's old age. The old vitality, not surprisingly, is ebbing.

Yet there is something almost hysterically desperate in Shaw's creation of these chosen people, as if he knew that humanity would not have time to wait for the evolution he had dramatized in *Back to Methuselah,* and would have to depend somehow for the immediate present on strong leaders who were considerably less than supermen.

The last play to have a successful production record was *The Millionairess,* written in 1944. Its success, however, was not assured until two years after Shaw's death when Katherine Hepburn played a vigorous Epifania in both London and New York. The story is made out of people who might have been left over from *Heartbreak House* — that is, useless, fashionable people who cannot handle their own affairs. But the Life Force has pointed its finger at two persons gifted with special vitality who might, together, provide a new kind of leadership. They are Epifania Fitzfassenden, a spoiled millionairess adventuress, and the penniless Egyptian doctor, who lives a life of service to science, humanity, and Allah.

Epifania's power does not stem from her money alone. "Shorn of her millions," Shaw says in the Preface, she would "still be as dominant as Saint Joan, Saint Clare, and Saint Teresa" (VI, 880). The lawyer in the play confirms this: "It's her will paralyses mine. It's a sort of genius some people have" (VI, 915). But Epifania is arrogant and selfish. The power she wields is not necessarily for good. The Egyptian doctor, who also has a powerful will and a dominant charisma, has no thought for money or material possessions. Epifania has promised her dying father that she would marry only a man who could turn a hundred and fifty pounds into fifty thousand within six months. She has already married a handsome but unremarkable athlete who has managed to pass the test, but he is nevertheless unsatisfactory and she is in the process of getting rid of him. She is now mightily attracted to the Doctor, but he, too, has made a deathbed promise — to his mother — that he would marry only a woman who could go out into the world with nothing but the clothes on her back and two hundred piastres (about thirty-five shillings, or, perhaps, five dollars) and earn her living alone and unaided for six months.

Both prove equal to their tests, but the Doctor uses his resources to rescue the widow of his former teacher from destitution and has no money left for himself. Epifania within a few months becomes the proprietress of the hotel where she began as a scullery maid. Like most of Shaw's unions, this one does not bode well for a peaceful marital home life.

> THE DOCTOR. . . . Shall I, the healer, the helper, the guardian of life and the counsellor of health, unite with the exploiter of misery?
> EPIFANIA. I have to take the world as I find it.
> THE DOCTOR. The wrath of Allah shall overtake those who leave the world no better than they found it.
> EPIFANIA. I think Allah loves those who make money. [VI, 962]

But Epifania consents to the marriage because you must learn to take chances in this world. And the Doctor finds Epifania's pulse irresistible.

The same theme is repeated in *Buoyant Billions,* written after Shaw had turned ninety. Here again the young man has to confess: "The Life Force has got me. I can make no conditions" (VII, 363). Until it is nearly time for the unlikely couple to apply for the marriage license, they do not even know each other's names, although the boy has pursued the girl halfway around the world. Both have had unusually permissive upbringing, and neither is equipped to deal with the hard problems of the world, although the young man presumes to be a "world-betterer" — "a missionary without an endowed established Church" (VII, 321) — and the young lady has posed as a kind of high priestess among Central American natives. Nevertheless it seems foreordained that, like Epifania, they must take their chances. Marriage, as the characters in this play discuss it, may mean many things. The interest of the Life Force is only in one: child bearing — "the replacing of the dead by the living" (VII, 355).

In his rambling but informative Preface to *Farfetched Fables,* which followed *Buoyant Billions,* Shaw maintains that the Life Force produces craftsmen, mechanics, legislators, thinkers, and geniuses in proportion as they are needed. It makes mistakes, of course, since it is neither omniscient nor omnipotent, and "proceeds experimentally by Trial-and-Error, and never achieves 100 percent success." He accepts, at age ninety-three, that he himself may be "one of its complete failures" (VII, 386).

> The Life Force, when it gives some needed extraordinary quality to some individual, does not bother about his or her morals. It may even, when some feat is required which a human being can perform only after drinking a pint of brandy, make him a dipsomaniac, like Edmund Kean, Robson, and Dickens on his last American tour. Or, needing a woman capable of bearing first rate children, it may endow her with enchanting sexual attraction yet leave her destitute of the qualities that make married life with her bearable. Apparently its aim is always the attainment of power over circumstances and matter through science, and is to this extent benevolent; but outside this bias it is quite unscrupulous, and lets its agents be equally so. [VII, 387]

The six Fables themselves are a return to the long-range dreaming of "As Far As Thought Can Reach" in which the Ancient Ones were about to become "vortices of pure thought." It occurs to Shaw now, in "the queer second sight," that perhaps such vortices are already in existence — if not from our own planet, then from elsewhere in the universe. He pictures one such vortex reassuming palpable form in the final Fable. Among the last fading images that Shaw put down on paper is a feathered creature who calls himself Raphael. He (or it?) materializes for a few moments in a schoolroom of the future, and thus, considerably before the universal UFO delirium, provides a "close encounter of the third kind"! The six Fables are loosely connected glimpses into the future that add nothing to the evolutionary ideas developed in *Back to Methusaleh*, and it is difficult to consider them as dramatic pieces for the stage. The first two assume a world where all the governments have atomic weapons, but everyone is afraid to use them. A chemist thereupon devises a lighter-than-air gas that will destroy life but leave property intact — a reasonably analogous prediction of the neutron bomb. The chemist sells his invention to "the South African Negro Hitler" and apparently to too many other heads of state as well, for he himself falls victim to the new gas, along with most of Western civilization.

The need for rapid evolution of creatures who could govern themselves concerned Shaw to his last breath. His final dramatic work was sent to the printers two months before his death. He called it *Why She Would Not*, and labeled it "a little comedy." He gave the appropriately named Bossborn the same natural power to rule with which he had endowed Epifania in *The Millionairess*. The characters are under-developed. The play would not last more than a half hour if performed. But it is not dull. For any Shavian the message has been overworked to the point of dreariness, but the language is still trenchant. In a sequence that Shaw had for some reason deleted from the version sent to the printer, he has Bossborn tell the heiress Serafina, who has been resisting his advances:

> You are not the only woman in the world, nor I the only man. Nature will still torment us with its demand for more children. I may come across a woman with whom I could not live for a single week. You may come across a man with whom the Life Force tells you you should mate, but with whom you could not talk for an hour without being bored beyond endurance. Yet your babies might be prodigies and mine geniuses. Nature does not care a rap for our happiness, only for our progeny. And, sex or no sex, we must leave the world better than we found it or this war-ravaged world will fall to pieces about our ears. [VII, 679]

But the final impulse was with Shaw the dramatist rather than with Shaw the apostle. Even in this distant echo of the artist in his prime, the people of his imagination have acquired wills of their own. Serafina knows that if she marries a Bossborn she will never be mistress in her own house, only his slave and his bedfellow; and this is "why she would not." She offers friendship only, and the Life Force must go searching elsewhere for the kind of union that will advance the race.

15

The Failure of Governments: *The Apple Cart*, *On the Rocks*, and *Geneva*

To anyone searching for some evidence that the Life Force was at work in providing improved leadership in government, the years between the wars were not encouraging. The failure of Versailles, the Great Depression, the rise of European dictatorships and their eventual demise in the cataclysm of World War II — these were the events that measured the period of Shaw's growing old. Throughout these years he struggled to hold on to his twin faiths in economics and religion. But it was now becoming ever more clear to him that little progress could be expected along either of these lines within the current political framework. No riddle was so agonizing for him as the Platonic one of what constitutes good government. That is perhaps why, until the very late years, all his plays skirt the problem, though clearly it is frequently on his mind.

In his non-dramatic writings, from the early *Essays in Fabian Socialism* to the later *Intelligent Woman's Guide to Socialism and Capitalism* and *Everybody's Political What's What?*,[1] and in countless shorter articles and speeches in between, Shaw's politics had turned out to be largely economics. But although universal economic well-being continued to be for Shaw the chief *purpose* of government, it did not determine a *method* of government; it merely made government more responsible and more difficult. With the full realization of his own waning powers, he finally forced himself to deal with the problem head-on in theatrical terms in *The Apple Cart*, *On the Rocks*, and *Geneva*. In the Preface to *The Apple Cart* (1929) he writes:

> We have to solve two inseparable main problems: the economic problem of how to produce and distribute our subsistence, and the political problem of how to select our rulers and prevent them from abusing their authority in their own interests or those of their class or religion. [VI, 254]

However, the most memorable moments in these three plays are those in which characters deliver bitter denunciations of government as Shaw saw it before the middle of the twentieth century. His complaints, as we shall see, are clear enough. His remedies are less so.

These are not among Shaw's best plays, and I'm sure he knew they were not, though two of them (I except *Geneva*) are still capable of holding and entertaining an audience. One reason why they are less satisfactory than *Man and Superman* or *Pygmalion* is that Shaw himself is floundering in dealing with the major thesis he has assigned himself. Strangely his most effective dramatic statement about government may be in *Heartbreak House*, where, ambiguous as it is, it is clear at least that the problem has to do with idleness and uselessness and drifting, and that the solution is "navigation." But the play does not dwell on who is to navigate or how the navigator shall be selected. The suggestion is merely made by Lady Utterword that her husband, Hastings, an effective governor of native peoples, could be the salvation of the country:

> LADY UTTERWORD. . . . Get rid of your ridiculous sham democracy; and give Hastings the necessary powers, and a good supply of bamboo to bring the British native to his senses: he will save the country with the greatest of ease.

And the old Captain responds, "It had better be lost" (V, 165).

But Hastings never appears in the play, and the subject of government becomes only a part of a pattern woven through one of Shaw's most complex tapestries. Even the three political plays of his old age do not yield many answers. Taken together, however, they do give insight into Shaw's most frustrating dilemma.

Shaw set *The Apple Cart* in what for him was the future, but what would be for us approximately the present. (One of the King's secretaries refers to the death of his father, presumably not very recent, in 1962.) As we observed in "The Tragedy of an Elderly Gentleman," Shaw's view of the future was not on all points sufficiently prescient; but in picturing western governments as caught between the forces of big business and big labor, his prophecy is accurate enough. The England of the play is completely dominated by "Breakages Limited," a multinational corporation, and there seems little hope of loosening its grip. The trade unions, too, in the person of Boanerges, are asserting their political powers. The crisis in leadership between the Prime Minister, Proteus, and King Magnus is principally a matter of who can best mediate among the forces that already exist. There is no indication that Breakages Limited can really be contained; it can only be handled diplomatically.

The problem in the play is complicated in the last act by the American

Ambassador's announcement that the United States has decided to rejoin the British Empire. "The Declaration of Independence is cancelled. The treaties which endorsed it are torn up" (VI, 355). Besides furnishing comic relief, the appearance of Ambassador Vanhattan recalls for us Clarence Streit's "Union Now" movement on the eve of World War II (only a decade after the play was written) when the proposition did not seem quite so preposterous. But in the play it gives King Magnus the opportunity for seeing through Vanhattan's offer and realizing that England would at that point merely be another star in the American flag, and that Breakages Limited would still be the real government.

In any case the American action is not at the heart of the play. The real upsetting of the apple cart is King Magnus's threat to abdicate and to stand for a seat in Commons. Since he is both more popular and abler than anyone in his cabinet, it is immediately apparent to the Prime Minister that in such a move his own power would come to an abrupt end. The "crisis" which Proteus had contrived is therefore called off, and the play is allowed to end with the ablest navigator still at the helm. But this is a theatrical expedient only. The issues that the play has presented — how to govern people who do not want to be governed, how to maintain the illusion of self-government when the real power is big business — are as unresolved at the end as they were at the beginning.

Though the time for *On the Rocks* is labeled "the present" (1933) and though it does indeed deal with the masses of the unemployed rebelling against a do-nothing plutocracy, it has, like *The Apple Cart*, a futuristic atmosphere, a detachment that allows the author to make explorations not hampered too much by contemporary facts. The Prime Minister here, Sir Arthur Chavender, is not as clever as King Magnus. Held almost captive by the angry crowds that surround Number 10 Downing Street, he has no idea how to handle the situation except by making speeches. But his speech-making practice is interrupted by a disputatious deputation from the Isle of Cats, which includes an old Laborite named Hipney. Chavender cannot, of course, satisfy the deputation's demands, and the members leave disgruntled, except for Hipney, who manages to suggest that the PM read Karl Marx. Later Chavender, distraught at his own inability to handle either the government or the affairs of his own family, is frightened by a mysterious lady psychiatrist, who nevertheless convinces him that he should spend some time at her retreat in Wales.

Rested, and forced to think, and full of Marxian readings, the PM returns in a few weeks to stage his own upsetting of the apple cart. He has made a vigorous socialistic speech at Guildhall, which brings his cabinet rushing to him. The conservative Sir Dexter Rightside is furious, but others, surprisingly, are favorable. They are not really in favor of socialism, but they see the PM's new stance as a political ploy which they can manipulate to their own advantage. As in *The Apple Cart* it is not expected that the real power will change, even though the political surface may seem to shift. When challenged by Rightside, Chavender must admit,

> I dont believe in the Class War any more than you do, Dexy. I know that half the working class is slaving away to pile up riches, only to be smoked out like a hive of bees and plundered of everything but a bare living by our class. But what is the other half doing? Living on the plunder at second hand. Plundering the plunderers. . . . No: there is no class war: the working class is hopelessly divided against itself. But I will tell you what there is. There is a gulf between Dexy's view of the world and mine. There is the eternal war between those who are in the world for what they can get out of it and those who are in the world to make it a better place for everybody to live in. [VI, 711]

But Chavender's hope for victory is short-lived. Though some of the politicians are willing to play along with him, the labor deputation from the Isle of Cats will not — because his program calls for compulsory labor and the abolition of strikes. On the second visit of the deputation, old Hipney again remains behind and delivers one of Shaw's most impassioned pleas for reform:

> Democracy was a great thing when I was young and we had no votes. We talked about public opinion and what the British people would stand and what they wouldnt stand. And it had weight, I tell you, sir: it held Governments in check; it frightened the stoutest of the tyrants and the bosses and the police: it brought a real reverence into the voices of great orators like Bright and Gladstone. But that was when it was a dream and a vision, a hope and a faith and a promise. It lasted until they dragged it down to earth, as you might say, and made it a reality by giving everybody votes. . . . That was the end of democracy for me; though there was no man alive that hoped as much from it, nor spoke deeper from his heart about all the good things that would happen when the people came to their own and had votes like the gentry. Adult suffrage: that was what was to save us all. My God! It delivered us into the hands of our spoilers and oppressors, bound hand and foot by our own folly and ignorance. It took the heart out of old Hipney; and now I'm for any Napoleon or Mussolini or Lenin or Chavender that has the stuff in him to take both the people and the spoilers and oppressors by the scruffs of their necks and just sling them into the way they should go with as many kicks as may be needful to make a thorough job of it. [VI, 718–19]

Hipney is not to be frightened by the word "dictator."

> Me and my like has been dictated to all our lives by swine that have nothing but a snout for money, and think the world is coming to an end if anybody

> but themselves is given the power to do anything.... Let [the people] have a voice. Let em have a choice.... Give em a choice between qualified men: there's always more than one pebble on the beach; but let them be qualified men and not windbags and movie stars and soldiers and rich swankers and lawyers on the make.... If you want to be a leader of the people, Srarthur, youve got to elect yourself by giving us a lead. Old Hipney will follow anyone that will give him a good lead; and to blazes with your elections and your Constitution and your Democracy and all the rest of it! [VI, 720]

It is not hard to imagine that in 1933 that speech brought cheers. But Chavender cannot meet Hipeny's challenge of leadership. After Hipney's departure Chavender decides to resign. His daughter and son have both found mates during the course of the play and are involved in the social melee. Outside the unemployed are again rioting, breaking windows, and singing "England Arise." "Suppose England really did arise!" Sir Arthur exclaims. It is the last line in the play, but Shaw adds a stage direction:

> *Unemployed England, however, can do nothing but continue to sing, as best it can to a precussion accompaniment of baton thwacks, Edward Carpenter's verses.* [VI, 736]

Shaw's hope for satisfactory government within a democratic framework is even dimmer than it was at the close of *The Apple Cart* four years before.

By 1938 democracy in Europe was a shambles, and it appeared to many that the era of representative government was nearing an end. Its apparent successor was dictatorship of some sort, but it could not be determined whether it would be of the left or the right — whether it would be Marxian and international in its outlook, or highly national and racial. Strangely Shaw saw some hope in this situation, and contributed his most topical play, *Geneva*, to the general discussion.

The flimsiest of plots serves as an excuse to bring caricatures of the dictators on stage in the last act for a lengthy debate on government. Mussolini (Bombardone), Hitler (Battler), and Franco (General Flanco de Fortinbras) respond to summonses of the World Court after complaints have been filed against them in the office of the forgotten International Committee for Intellectual Cooperation. Significantly Stalin is not called upon, although in the first act a complaint is lodged against the Soviet state that is hardly more fanciful than those which have brought the other dictators before the international tribunal. But the plaintiff against Russia is an English bishop who has discovered that his footman is a Communist organizer. In the Office of the International Committee for Intellectual Cooperation, however, the Bishop confronts a Russian Commissar, Posky, who counters each of the Bishop's charges with his own accusations against Christian missionaries' attempts to subvert the Russians. The shock of some of Posky's statements is too much for the elderly Bishop, twice causing him to faint. And when Posky assures him that there are no poor in the Soviet Union, the old gentleman simply drops dead (as well he might!). The plaintiff thus disposed of, the Russian dictator is not called upon to join his western peers in the last act. But Commissar Posky remains in the play to represent Socialist Russia and to remind the others how the problems of government really ought to be handled.

It is worth recalling here that Stalin was the only one of the dictators whom Shaw had met personally. In 1931 he spent nine days in the Soviet Union as part of a group of five which included Lord and Lady Astor, their son David, and Philip Kerr — the future Lord Lothian. The Russians made much of Shaw as the famous English Marxist. He and the Astors were granted an interview with Stalin which lasted an unprecedented two hours and thirty-five minutes. The visitors apparently did not spare Stalin some criticism of the Soviet system, but at the close of the session Shaw was full of praise for the Russian dictator. He found "no malice in him, but also no credulity."[2]

It may be said that there is no malice either in Shaw's pictures of the other dictators, but there may be some measure of credulity. Though he makes them appear vain and foolish, he gives them ample reasons for their dictatorial methods. One gathers that all three have improved the lives of their countrymen. Bombardone, for example, cannot be bothered with the charge that he has destroyed liberty and democracy. "My business is government," he says. "I give my people good government as far as their folly and ignorance permit. What more do they need?" (VII, 133). And Battler claims, "I have made better men and women of [my countrymen]. I live for nothing else. I found them defeated, humiliated, the doormats of Europe. They now hold their heads up with the proudest; and it is I, Battler, who have raised them to spit in the faces of their oppressors" (VII, 140).

Flanco, when his turn comes, says, "I stand simply for government by gentlemen against government by cads. I stand for the religion of gentlemen against the religion of cads. For me there are only two classes, gentlemen and cads: only two faiths: Catholics and heretics. The horrible vulgarity called democracy has given political power to the cads and the heretics. I am determined that the world shall not be ruled by cads, nor its children brought up as heretics" (VII, 148–49).

Shaw's note in the Malvern Festival Book, 1938, concludes: ". . . Instead of making the worst of all dictators, which only drives them out of the League, I have made the best of them, and even have given them some measure of fair play. I hope they will like it" (VII, 167). In the program for the first production of *Geneva* in London he added a challenge to the dictators "to live up to those portraits if they can."[3]

There are two other characters in *Geneva* who provide some insight into Shaw's attitude toward government. One is Sir Orpheus Midlander, the British Foreign Secretary, who is the satirical epitome of British complacency and muddle-through. A sample of his rhetoric will be sufficient:

> When you ask me what will happen if British interests are seriously menaced you ask me to ford the stream before we come to it. We never do that in England. But when we come to the stream we ford it or bridge it or dry it up. And when we have done that it is too late to think about it. We have found that we can get on without thinking. You see, thinking is very little use unless you know the facts. And we never do know the political facts until twenty years after. Sometimes a hundred and fifty. [VII, 123]

The other character worth note is Begonia Brown, the typist at the International Committee for Intellectual Cooperation, who, in the first act, innocently refers all the complaints to the Court of International Justice and thus causes the play to happen. Her service to the plot is finished at that point, but Shaw retains her as the lowest common denominator of universal suffrage. She is intensely loyal to Camberwell, a jingoist, vigorously opposed to communism and the Bolshies, but almost as virulently opposed to Peckham — Camberwell's rival district in London. She never reads the political news and is "a complete ignoramus"; but she is attractive and assertive, and on the basis of her new-found notoriety she is elected to Parliament and is made a Dame of the British Empire. Later, in writing about the play, Shaw charges that "giving the casting vote to Miss Begonia Brown" was "a guarantee of petty snobbery and parochial interest in the choice of rulers" (VII, 173). Echoing old Hipney in his earlier play, he tells us that "Democracy did great things when it was an ideal. It was its reduction to reality in the idiocy of Begonia Brown that produced the snobocracy of the last twenty years" (VII, 176).

Clearly Shaw assumes that the most important ingredient of good government is competent leadership. He seems to imply again and again that if we could only be assured of leadership of intelligence and integrity — not Hastings Utterword's kind — the problem of government would be largely solved. But we are not really interested in *self*-government. "Our professed devotion to political principles," he says, "is only a mask for our idolatry of eminent persons" (VI, 250). We have observed Shaw's fascination with that personal phenomenon which we have since learned to call charisma, but for which he had no name. In his Preface to *The Millionairess* ("Preface on Bosses") he borrows Bulwer-Lytton's term, *Vril* (VI, 879), to identify the magic that somehow sways masses and makes obedient followers out of previously apathetic anarchists. Is this mysterious quality evidence of the Life Force flowing through these people? Do they find themselves possessed, as the long-livers did in *Back to Methuselah*, of special attributes that mark them apart from the common herd? Are they, whether they know it or not, part of the evolutionary thrust into the future? Shaw thought that they might be, even hoped they might be. Hence his fascination with historical figures such as Napoleon (*The Man of Destiny*), Julius Caesar (*Caesar and Cleopatra*), General Burgoyne (*The Devil's Disciple*), Catherine the Great (*Great Catherine*), Joan of Arc (*Saint Joan*) and Charles II (*"In Good King Charles's Golden Days"*); and such fictional leaders as Bluntschli (*Arms and the Man*), Lady Cicely Waynflete (*Captain Brassbound's Conversion*), Andrew Undershaft (*Major Barbara*), Tom Broadbent (*John Bull's Other Island*), King Magnus (*The Apple Cart*), Epifania Fitzfassenden (*The Millionairess*), extending even into the posthumous *Why She Would Not* with Bossborn.

Conversely his faith in democracy steadily declines in his later years. There can be little doubt that Hipney's disillusion with universal suffrage is in part Shaw's own. In 1888 he saw democracy as a welcome and necessary stage for the transition to socialism.[4] In 1910 he is of the opinion that equality of income must *precede* democracy. "Democracy would really work then, and would be irresistible and permanent, instead of being what it is at present, a doubtful and discredited recent political experiment, which has achieved its successes, such as they are, only because it is not really democracy at all."[5]

But by the end of World War I and thereafter, he was accepting democracy only in terms of his own definitions. "Democracy means government in the interest of everybody. It most emphatically does not mean government BY everybody" (VII, 407). On numerous occasions he took apart Lincoln's phrase to separate "Government of the people, and for the people" — propositions which he found essential — from "Government by the people" — which he found preposterous and dangerous.[6] The clever politicians dupe the people into thinking they are governing themselves, when in reality "the financier and the soldier are cocks of the walk; and democracy means that their parasites and worshippers carry all before them" (VI, 871). The art of government is far too difficult and specialized to be entrusted to Begonia Brown. Nevertheless Shaw was forced to accept (reluctantly, it seems to me) the proposition that government must be by *consent* of the governed. But this is further complicated by the observation that the governed really don't want to be governed. "Have I not indeed just been impressing on you," Shaw lectures in *The Intelligent Woman's Guide to Socialism and Capitalism*, "that the miseries of the world today are due in great part to our objection, not merely to bad government, but to being governed at all?"[7] Apparently as close as the governed ever get to "consent" is the acceptance of a commonly acknowledged leader of competence and integrity.

We have already noted that Shaw's thinking is essentially economic, and that his economics is essentially Marxist. The chief purpose of Shaw's government would be to move rapidly and efficiently toward a universal equality of income. His government would also, I presume, allow open criticism, and permit everyone the right to his or her individual morality, so long as it did not interfere with the rights of the community as a whole. Indeed, good government should set people free to make their own contributions, and not bother them with matters outside their competence.

Shaw's concern with rulers, and how they are to be chosen and controlled, skirts many of the essential elements of political economy. He contributes almost nothing toward the constructive reorganization of the actual machinery of government. He found the checks and balances written into the American constitution simply a hindrance to getting things done, a guarantee of anarchy.[8] He also advocated the scrapping of the British Parliament as obsolete. But government must be organized in some way, and what does Shaw recommend? He would replace Parliament (and, presumably, the American legislative bodies) with two or three federal legislatures organized along the lines of the British municipal committee system — and "with a central authority to coordinate the federal work" (VI, 871). Shaw, it must be remembered, was familiar with the municipal committee system through his six years of service as vestryman and borough councillor for the Borough of St. Pancras beginning in 1897. (He was appointed to the position by political friends.) The vestry consisted of 120 members, the council of seventy. Most of the work of these bodies was effected through small committees, and during his tenure Shaw served on committees dealing with electric lighting, health, relations with Parliament, housing, and drainage (sewage). Though he failed in such worthy efforts as getting free municipal toilets for women, he obviously felt that much good had been accomplished; and he was entirely willing to continue a minor career in politics by running for the London County Council. But he was, as he had suspected, simply unelectable.

The Borough of St. Pancras contained about 250,000 Londoners, and their problems of sewers and housing conditions, certainly of the greatest importance to the people involved, were a far cry from the problems of the national budget, employment, foreign trade, and peace and war that national governments must handle. If we are to escape the impression of total naiveté, we must assume that the "central authority" of which he speaks is indeed a *strong* central authority, and that the weight of government, for better or worse, must inevitably rest on the charismatic leader.

It would appear, then, that the most satisfactory form of government would be a benevolent and enlightened dictatorship. But the history of dictatorships does not give any clue as to how to assure benevolence or enlightenment in a ruler who no longer needs to fear any opposition. Throughout his career Shaw was fascinated by Napoleon. It is safe to say that he makes more references to him than to any other political figure. Shaw credits him with extraordinary military talents, gives him credit for restoring order in France and ruling more ably than the Directorate which he superseded. But he became "bankrupt as a glory merchant" and in the end it was "quite in order to call Napoleon a snob, a cad, an assassin, and a scoundrel."[9] The contemporary dictators (except for Stalin) do not fare any better. Ideally, Shaw thinks, a dictator, however powerful, should operate within a predictable framework that could be recognized as a sort of constitution. "Failing any ... religious or political creed all autocrats go more or less mad. That is a plain fact of political pathology" (VI, 871).

> Mussolini and Hitler have failed in this respect. They have won their eminence by doing things that everybody wanted done, and been idolized accordingly; but having no established and generally intelligible creed, no one can feel any assurance of where they will stop; and so, to keep their supremacy, they have to feed their idolators with military glories and their financiers with commercial successes, which is not continually possible.[10]

Mussolini, on the whole, gets higher praise than Hitler, whose theories of racial purity Shaw finds both absurd and repugnant. General Franco, by the way, except for his brief appearance as "Flanco" in *Geneva*, gets hardly a mention elsewhere in Shaw's writings.

Three years prior to *Geneva*, Shaw was defending Mussolini:

> The English editors ... wrote sympathetic articles paraphrasing John Stuart Mill's Essay on Liberty. Mussolini, now Il Duce, never looked round: He was busy sweeping up the elected municipalities and replacing them with efficient commissioners of his own choice, who had to do their job or get out. The editors had finally to accord him a sort of Pragmatic Sanction by an admission that his plan worked better than the old plan; but they were still blind to the fact staring them in the face that Il Duce, knowing what the people wanted and giving it to them, was responding to the real democratic urge whilst the cold tealeaves of the nineteenth century were making them sick. [VI, 864]

Hitler's foolish racism, on the other hand, reached its climax for Shaw with the Fuehrer's persecution of Einstein:

> Hitler's throwing Einstein to the Antisemite wolves was an appalling breach of cultural faith. It raises the question which is the root question of this preface: to wit, what safeguard have the weaponless great against the great who have myrmidons at their call? [VI, 866]

Do any of the dictators, present or past, get really good marks? Yes: Julius Caesar, Mahomet, Cromwell, and George Washington made "the best of their power or at least not the worst of it."[11] But "only Cromwell with his Bible and Convenants of Grace, and Stalin with his Marxist philosophy, held themselves within constitutional limits (as we say, had any principles); and they alone stand out as successful rulers."[12] These comments were made in 1944 when Shaw was eighty-eight. He was committed, of course, to the Russian experiment. One can only speculate what turn his disillusionment would have taken if he had lived long enough to realize Stalin's megalomania and cruelty.

Shaw's mistake in the case of Stalin underscores not only the danger of entrusting so much power to a single ruler, but also the difficulty of making a choice. Who is to do the choosing? Shaw was right in 1928 when he conceded the general population must somehow be represented in this process, and that there must be some established pattern for ruling and for succession. In other words, in spite of his later protestations to the contrary, there must be a parliament and a constitution.

> . . . dictatorial strong men soon die or lose their strength, and kings, generals, and proletarian dictators alike find that they cannot carry on for long without councils or parliaments of some sort, and that these will not work unless they are in some way representative of the public. . . .[13]

And how is the choosing to be done? Here Shaw contributes what may be his only creative and practical idea to the problem of government. There can never be any guarantee that the choosers — whether the general electorate or a parliament or a council — will select the best possible leader. But there ought to be some assurance that the choice can be made from among qualified individuals. (We recall old Hipney's plea, "Give em a choice between qualified men: there's always more than one pebble on the beach. . . .") Late in life Shaw developed an interest in what he called "anthropometry" — what we would call psychological testing. He devotes a rambling chapter to it (Chapter XXXVI) in *Everybody's Political What's What?* Of the civil service examinations and the kinds of examinations given for professional licensing he says that they exclude "thinkers whose memories will not retain things not worth remembering"; and of the psychological testing developed between World War I and World War II: "I have never yet come across an intelligence test that I could pass. . . ."[14] But he did not therefore condemn the process. Shaw sees in the process of testing some long-range hope for eliminating at least the most grossly incompetent of would-be governors. Obviously the testing procedure must be vastly improved over anything he knew of — or that we know of. Predictably, the content of the tests as he sees them would be largely economics: labor costs, production costs, theories of the distribution of wealth. But candidates might also be examined for creed and conduct and esthetic responses. Though we do not yet know how to do this, it is not too soon to make a start.

It must now be obvious that Shaw was sceptical of all forms of government, and was completely frustrated in his attempt to establish a clear guide-line through the maze, such as he found in his thinking about economics and religion. His mistrust of universal suffrage and his natural attraction to "great men, so called"[15] led him toward at least a temporary acceptance of dictatorship and complete acquiescence to the rule of Stalin. But the number and the magnitude of unanswered questions did not let him rest easily. His latter-day interest in anthropometry seems merely to indicate that we need to understand better the nature of man before we can improve the nature of government — a proposition which would be hard to refute. In the end he had to concede that government by the people and government by dictators "are impossible extremes, not inevitable alternatives."[16] To anyone searching for absolute answers, this is not very satisfying.

In his view of the future in *Back to Methuselah,* whole populations of longlivers had to be evolved before society learned to handle itself rationally and before the problem of government seemed to disappear. But there is no evidence that the Life Force can act rapidly enough to provide political solutions in our own times. To bring his discursive play, *Geneva,* to a close, and finally allow his audience to go home, the playwright had to invent news of a sidereal catastrophe: ". . . the orbit of the earth is jumping to its next quantum. . . . Humanity is doomed." The report proved to be false, but it put an end to the on-stage argument about government. Off-stage there was no resolution in sight.

16

Postlude with a Nun and *The Black Girl*

In *Back to Methuselah* Shaw presented the Life Force, or Creative Evolution, as a universal religion which would embrace all other religions. He was not, I am sure, proposing a new sectarian "church" — even though the West London Ethical Church had enshrined him in a stained glass window along with Anatole France and Saint Joan. His arguments remained on an abstract intellectual level. He seemed always more eager to convince seeking doubters among the scientific determinists than to convert orthodox Christians to Creative Evolution. But the Life Force was destined to learn just how disarming and loving — and unyielding — orthodox Christianity can be!

In April of 1924, when Shaw was already sixty-eight, he and his wife, Charlotte, first called on Dame Laurentia McLachlan, the Benedictine sister who was to become Abbess of Stanbrook Abbey near Worcester. They spoke with her through a grille in the guest parlor, for the Benedictines are an enclosed order with minute regulations governing all aspects of the nuns' lives. Normally an enclosed nun would spend her whole life within the precincts of her house without ever crossing the threshold of the great enclosure door. Nor could any outsider be admitted beyond the grille except for emergencies. Dame Laurentia, who became an authority on the Gregorian chant, was sent to give instruction at other houses, but for these excursions she had to obtain a papal dispensation. She had, as Shaw later observed, an "unenclosed mind," but through his visits and letters over the next twenty-five years more of the secular outside world found its way into Stanbrook than she might have bargained for.

The quarter-century friendship was of necessity mostly carried on by correspondence — often stormy. Fortunately most of Shaw's letters were preserved in the Abbey, and after the Abbess' death in 1953 the sisters included some of them in their book, *In a Great Tradition*, a life of Dame

Laurentia.[1] The existence of the letters had been known through the memoirs of Shaw's secretary, Blanche Patch,[2] and the sisters had allowed a prior publication of them in the July and August 1956 issues of the *Atlantic*. A few passages have been excised from the letters where "reverence forbids quotation, for the profanation of divine names and ideas becomes revolting and unbearable."[3] These passages will no doubt be restored in the forthcoming volumes of *Bernard Shaw, Collected Letters* being edited by Dan H. Laurence. Sometime after the publication of *In a Great Tradition* sixteen of Dame Laurentia's letters to Shaw, having been bequeathed to the British Museum under the terms of Shaw's will, became available. So we now have enough of the exchange for some insight into a fascinating episode.

Margaret McLachlan, from a Scottish Catholic home, the youngest of six children and admittedly spoiled, had seen little of the world before she entered the convent as a novitiate at eighteen and took the name of Laurentia. She had been briefly to school at Edinburgh and was later enrolled at Stanbrook. But most of her education, like Shaw's, had been informal. She was a delicate and precocious child, with a good library at her disposal, and an elder brother who was a priest and who enjoyed teaching her.

The first meeting of the Shaws with Sister Laurentia had been prepared by their mutual friend, Sir Sydney Cockerell. His association with Stanbrook went back to 1907, and for all Laurentia's enclosed life it formed a window through which she looked out on the non-Catholic world. Sir Sydney was an agnostic humanist, a disciple of Ruskin and a friend of William Morris. The unorthodoxy of both Cockerell and Shaw was to cause her heartache. Of the two, Cockerell remained the more conventional agnostic. He tried to interest her in Tolstoy (who disturbed her), and she referred him to the *Confessions* of Saint Augustine and the Penny Catechism. He raised the standard scientific-humanist objections to the Ascension, the Resurrection, Hell, and the doctrine of Infallibility. She, with loving patience and refreshing humor, tried to show him that these dogmas set the searching mind free. One cannot help feeling that she had the better of it, that her Catholicism was better grounded than his agnosticism. Though she could not convince his intellect, the richness of her simple life makes his studied unbelief seem almost an academic pose.

In Shaw, however, she faced not so much an unbeliever as a man with a rival religion. The nun, ten years Shaw's junior, knew little of Shaw's work before Cockerell gave her a prepublication copy of *Saint Joan*. This she thought "a wonderful play" though she took exception to some of the aspects of Joan's trial. Shaw found great delight in the fineness of her mind, and their extended arguments about Catholicism brought, for a time, obvious enjoyment to them both.

Charlotte initiated the correspondence and arranged for the 24 April meeting. The Shaws talked pleasantly through the bars with Laurentia and discussed *Saint Joan*: "Mr. Shaw gave good reasons for his treatment of *Saint Joan* with regard to the points which I questioned," she reported to Cockerell. But apparently Cockerell later gave her an ambivalent report of the meeting:

> S.C.C. told me that after Shaw's first visit he received an enthusiastic report of it, and he asked Shaw when he was coming again. "Never," said G.B.S. . . . Then he reflected and asked, "How long has she been there?" "Nearly fifty years." "Oh, that alters the case, I'll go whenever I can."

Laurentia thought that the conversation indicated "that the life here, and therefore the Church, does attract him," and she prayed for "grace to help this poor wanderer so richly gifted by you" (*IGT*, 236).

The Shaws visited again during the summer. Shaw purchased *The Letters of St. Teresa of Avila*, recently translated and printed at Stanbrook. On 1 October he inscribed a copy of *Saint Joan* "To Sister Laurentia from Brother Bernard." It was her first look at the Preface. She was pleased, but had a few demurrers. She was in no way deferential to him. "When he next shakes my bars I shall make some further remarks," she wrote Cockerell.

It was Shaw's Christmas letter (23 December 1924) that really began their long theological debate, which continued with some interruptions until his death twenty-six years later. She had sent him a paragraph from the *Book of Wisdom* — the portion selected for the Epistle of the Mass appointed for the Feast of Saint Joan — to show that the attitude of the Church was one that honored Joan's special insight. She also had sent him *The Godly Instructions and Prayers of Blessed Thomas More Written in the Tower of London 1535*. The letter that accompanied these gifts is lost, but we can surmise its contents from Shaw's reply.

He explained that his own writing was for a wider public than just Catholics or just Christians, and that God allows different people their own iconographies.

> Christ in his metaphor of the tares and the wheat, has given us a very plain warning to let Allah and Brahma and Vishnu alone, as if our rash missionaries pluck them out of the Arab and Hindu soul, they will pluck all the religion out of that soul as well. . . . It was therefore necessary for me to present Joan's visions in such a way as to make them completely independent of the iconography attached to her religion.

And the visions remain miraculous, whether the messengers are considered to be real persons or inspired illusions.

Shaw reminds Laurentia that there is a strong element of rationalism running through Catholicism, but for both Catholics and non-Catholics rationalism is not sufficient. And it is here that he dates his own abandonment of rationalism and his reliance on mysticism as 1879, the year he completed his second novel. (See page 19).

As to the reading material she had sent him — he was "amazed and delighted" with the passage from the *Book of Wisdom*, but he found it difficult to accept much of Thomas More. He found "the Psalm part of it a mere literary exercise," and thought Psalms generally were "classic examples of fool's comfort."

> Comforting people by telling them what they would like to believe when both parties know that it is not true is sometimes humane and always to be let off with a light pennance [sic] as between two frail mortals; but it should not be admitted to the canon.

Even here at the beginning of the argument it is clear that Shaw is somewhat torn between his desire to be fully candid and the need to be gentle in his persuasion. "I must stop, or I shall begin by kicking my cloven hoof too obviously for your dignity and peace; but I mean well, and find great solace in writing to you instead of to all the worldly people whose letters are howling to be answered" (*IGT* 241).

I have no letters between this one and 1930, but it is evident that Laurentia presented Shaw with a unique dilemma. Shaw carried on a voluminous correspondence with all sorts of people. His letters are often brilliant (he seemed always to be writing for a larger audience), but he let his guard down with very few. Now here was a woman with whom he might really be able to communicate, someone who seemed to understand him at a very deep level. Cockerell, indeed, reported that he "never saw [Shaw] so abashed by anyone but William Morris," and that he "seemed to admit that he was in the presence of a being superior to himself." (*IGT*, 242–43). Yet she was not a woman of the world — he could not discuss politics or the London theatre with her. The subject matter must inevitably be, no matter how it was styled, religion. And how long could this world-famous heretic and iconoclast remain within the affectionate embrace of this daughter of the Church, who was, though liberal in outlook, essentially and devoutly orthodox?

Shaw used the Christian imagery skillfully, always retaining his own interpretations. The Benedictine writers of the book note that Shaw had "an extraordinary feeling for Christian tradition," and this is true, but deceptive. Shaw's conception of the Holy Ghost, for instance, was that of an impersonal Life Force, and far removed from that of a good Catholic; yet he uses the term freely throughout his later work.

Sister Laurentia was able to respond to Shaw's iconoclasm with patience and restraint. She was able to perceive the essential simplicity and humility of "Brother Bernard." They had other things in common: music, although strangely they did not discuss it, even though one was an accomplished music critic and the other the leading authority on the Gregorian chant. They both were disciplined ascetics. And humor. Margaret McLachlan chose the religious name of Laurentia after Saint Laurence the martyr partly because he could joke while he was being tortured! ("Stripped and bound to a gridiron to roast over a slow fire, he mocked his tormentors. 'Turn me over. I'm done on this side . . .'" [*IGT*, 95].) The humorist of Ayot St. Lawrence too, had learned to laugh while he was being roasted, as he explains in his 1921 Preface to *Immaturity*. ("If you cannot get rid of the family skeleton, you may as well make it dance.")[4] But it is not likely that the nun realized that she was being drawn into so fundamental a conflict.

One day in December of 1930 Shaw called but found her too ill to see him. He was genuinely distressed and wrote at once to say that he had telepathically divined that something was wrong, and to assure her that "whatever corresponds in a heathen like me to prayers" had been sent out in her direction. Laurentia recovered, but remained in delicate health.

In the following March the Shaws set out on a tour of the Mideast with a party that included Dean Ralph Inge, the controversial Anglican Dean of St. Paul's. She wrote Shaw: "Before you start for the Holy Land I want to send you my blessing and ask you to take me, my spirit, with you, and let it run about in the Holy Places which I know without having seen them." And she asked him to bring her back "some little trifle from Calvary." By this time she was addressing him "Brother Bernard" and signing herself "Sister Laurentia."[5]

Shaw took her request seriously. On the 15th he sent her a hurried note as they were leaving Jerusalem for Nazareth, promising to write her more fully when he got back to ship. His long descriptive letter was begun on St. Patrick's Day, continued on shipboard as they sailed for Cyprus and Rhodes, but not completed until the 26th. The Sister had every reason to be pleased with it. It is one of Shaw's best letters, both for its keen observation and for its personal charm. His easy and accurate references to the Bible and to early Church history made him an ideal tourist's guide. They arrived in the Holy Land at night, and his first view of it was at dawn: "On this first hour you do not improve. It gives you the feeling that here Christ lived and grew up, and that here Mary bore him and reared him, and that there is no land on earth quite like it."

But when the guides start pointing out the spots where Jesus did this or that, Shaw is as repelled as Laurentia would also have been. She would certainly have endorsed his view that "it is better to have Christ everywhere than somewhere, especially somewhere where he probably wasn't." Shaw managed to send her a few olive leaves from the Mount of Olives, but the "trifle" from Calvary he could not provide, since he was convinced that no one knew which of the many hills surrounding Jerusalem was the true one.

Instead he is bringing her, he says, two little stones from Bethlehem — "a scrap of the limestone rock which certainly existed when the feet of Jesus pattered about on it and the feet of Mary pursued him to keep him in order" (*IGT*, 245–51). One of the stones is for Laurentia personally. The other is to be thrown blindfold into the Stanbrook garden, so there will always be a stone from Bethlehem there, but no one will know which one it is and be tempted to steal it. (Laurentia did not remind him, as she might have, that the Holy Family probably left Bethlehem before Jesus was old enough to do any pattering about.) The entire letter was calculated to please her. It maintained a light tone and a sceptical attitude without ever being really irreverent. Laurentia could hardly have objected to anything except the closing sentences, where Shaw ascribes the Book of Revelation to an obvious drug addict.

Even to this, however, she responded with good humor: "It is well for you that you did not live in the Middle Ages. You would have been boiled in a cauldron of oil for your remarks on the Apocalypse!" Otherwise her reply (beginning on the 18th of April, but not finished until the 21st) was all affection and gratitude. "You have made me feel that I have seen the Holy Land through your eyes, and have revealed a great deal more than I should have seen with my own." It is well, she agrees, that Christ cannot be located in Palestine — that there is no Christian Mecca. But "this childish mania for labeling everything seems to be universal." Part of Catholicism's poor showing in Palestine she blames on the fact that the Church "is largely represented by Franciscans who are more devotional than liturgical, and who do not abound in taste."[6]

On 21 April she added: "Your letter has come with the little souvenir of St. Francis." We do not have Shaw's letter that accompanied this additional gift, nor do we know what "the little souvenir" was. The Bethlehem stone for the garden arrived on the 12th of June. Shaw wished to deliver the other one personally. On his return, however, he found that Laurentia was seriously ill, probably in part from overwork, and had been ordered to bed by her doctor. In the meantime he took advantage of an unexpected opportunity to visit the Soviet Union for nine days, where he spent his seventy-fifth birthday and had his much-publicized interview with Stalin. ". . . the oddest place you can imagine," he wrote her. "They have thrown God out the door; and he has come in again by all the windows in the shape of the most tremendous Catholicism" (*IGT*, 251).

There was a further reason for the delay of the delivery of the second stone. It was not until the 18th of September, when the Shaws were again in Malvern for the Festival, that they asked if they could be received at the Abbey. On the following Saturday they appeared with the long-promised icon. It was more of a presentation than anyone at the Abbey had anticipated. The Shaws had had the silversmith, Paul Cooper, set the stone in a silver reliquary surmounted by the figure of the infant Jesus, his left hand supporting the globe, his right raised in blessing. Most of the outside world would have been even more surprised than the Benedictines at this Shavian display of Christian iconography. One can only assume that it was Charlotte with whom Cooper had consulted on the matter of the design. But for Laurentia this must have been one more piece of evidence (along with the wonderful letter from the Holy Land, the olive leaves, and the souvenir of St. Francis) that GBS was really a devout Christian who had not yet sufficiently conquered his pride to make a confession. Apparently she was not aware of Shaw's earlier pronouncements on religion, such as his prefaces to *Androcles of the Lion* and *Back to Methuselah.* Though she would have been perfectly free to find out more about the religious views of her confidant than what he told her in his letters (most literate readers of the English language already knew them), the fact of the moment was that she was surprisingly innocent. Shaw in his letters was careful, as always, to say nothing that he could not in some sense believe; yet we must hold him responsible for a kind of protective duplicity. He obviously wanted to retain this friendship as long as he could. Perhaps Laurentia, too, sometimes purposely looked the other way. Inevitably she would have to confront the real Shavian heresy, and she was not prepared for it.

At this moment, however, she and her sisters were simply carried away with the beauty of the gift, and she warned him that "by making such a gift to a place like this you expose yourself to the danger of being prayed for very earnestly. . . . You have made yourself our debtor [sic] & must take the consequences."[7]

Shaw replied that he did not in the least mind being prayed for. He thought of prayers as wireless messages floating around in the ether. "I suppose if I were God I could tune into them all. . . . If the ether is full of impulses of good will to me so much the better . . ." (*IGT*, 253).

Sir Sydney Cockerell, who dropped by to admire Cooper's artistry, suggested that the reliquary should have an inscription saying that it was from Bernard Shaw to Sister Laurentia, and explaining its purpose. Laurentia passed the suggestion along to Shaw, whose response must certainly rank with the most disarming ever penned:

> Dear Sister Laurentia
>
> . . . Why can it not be a secret between us and Our Lady and her little boy?

What the devil — saving your cloth — could we put on it?

Cockerell writes a good hand. Get him a nice bit of parchment and let him inscribe it with a record of the circumstances for the Abbey archives, if he must provide gossip for antiquarian posterity.

We couldn't put our names on it — could we? It seems to me something perfectly awful.

"An inscription explaining its purpose"! If we could explain its purpose we could explain the universe. I couldn't. Could you? If Cockerell thinks he can . . . let him try, and submit the result to the Pope.

Dear Sister: our finger prints are on it, and Heaven knows whose footprints may be on the stone. Isn't that enough?

Or am I all wrong about it?

faithfully and fraternally
Brother Bernard [*IGT*, 252–53]

One hardly needs to report that the reliquary remains uninscribed.

At the end of October her health is still not good. But the prayers for Brother Bernard's soul continue. "Do you know I began to pray for you before I ever saw your face? I then called you (to the Lord) 'Bernard Shaw,' but now you are 'Brother Bernard.'"[8]

Now there is a break in the correspondence, and we must piece together the sequence of events as best we can. As of January 1932, Sister Laurentia became the Mother Superior — the "Lady Abbess" of Stanbrook. Also at the beginning of 1932, the Shaws set off on another trip, this time to South Africa. Shaw claimed that he made these trips only to indulge Charlotte, that he did not enjoy them himself; but once on the journey — as is obvious from his letters from the Holy Land and Russia — he developed all the symptoms of an enthusiastic tourist. The South African sojourn, however, was marred by a serious accident.

Shaw, with Charlotte and another companion, was driving a rented car to Port Elizabeth. Shaw was not known as a good driver — he had had a number of minor accidents with vehicles. In this case he mistook the accelerator for the brake, and plunged off the road and over a bank. He and his companion were uninjured, but Charlotte, who was in the back seat with the luggage, was badly bruised, her shin punctured, and her wrist sprained. It took more than a month for her recuperation.

Shaw used this time to write, not a play, but a fable modeled on the form of Voltaire's *Candide. The Adventures of the Black Girl in Her Search for God* follows the path of a simple but highly curious native girl who has been told by the missionaries, "Seek and ye shall find." She quickly finds, and destroys with her knobkerry, the savage gods of Moses and Job, and pauses not much longer for the despairing Ecclesiastes or the howling prophet, Micah. The African forest in which she searches is not entirely chronological, and in flight from the prophet she comes upon a more modern god,

Science, in the obvious guise of Ivan Pavlov, whose conditioned reflexes land him in a tree where he is frightened by an imaginary crocodile. By a well she meets also the kindly Jesus (called "the conjurer") to whom she is attracted. She leaves him with reluctance when she finds that even he cannot give her satisfactory answers. But after further adventures with a caravan of white explorers, she comes upon the conjurer again. This time he is not alone.

He is lying on the ground, stretched out on a large wooden cross, while an artist, who specializes in gods and goddesses, carves a statue of him. While this is going on, an "Arab gentleman" (Mohammed) sits nearby and argues with the conjurer about the necessity for violence in government, the compatibility of art and religion, and polygamy. The conjurer complains:

> "I am so utterly rejected of men that my only means of livelihood is to sit as a model to this compassionate artist who pays me sixpence an hour for stretching myself on this cross all day. He himself lives by selling images of me in this ridiculous position. People idolize me as the Dying Malefactor because they are interested in nothing but police news. When he has laid in a sufficient stock of images, and I have saved a sufficient number of sixpences, I take a holiday and go about giving people good advice and telling them wholesome truths. If they would only listen to me they would be ever so much happier and better. But they refuse to believe me unless I do conjuring tricks for them; and when I do them they only throw me coppers and sometimes tickeys, and say what a wonderful man I am, and that there has been nobody like me ever on earth; but they go on being foolish and wicked and cruel all the same. It makes me feel that God has forsaken me sometimes."[9]

Weary of searching, the Black Girl discovers at last the wise old gentleman (Voltaire) who advises her that the best place to seek God is in a garden. "You can dig for Him here." "And shall we never be able to bear His full presence?" she asks. "I trust not," he answers. "For we shall never be able to bear His full presence until we have fulfilled all His purposes and become gods ourselves." So the Black Girl settles down to work with the philosopher. They are joined at length by a red-haired Irishman, a Socialist who is not interested in searching for God.

> "Sure God can search for me if he wants me. My own belief is that he's not all he sets up to be. He's not properly made and finished yet. There's somethin' in us that's dhrivin at him, and somethin out of us that's dhrivin at him: that's certain; and the only other thing that's certain is that the somethin makes plenty of mistakes in thryin to get there. We'v got to find out its way for it as best we can, you and I; for there's a hell of a lot of other people thinkin of nothin but their own bellies." And he spat on his hands and went on digging.[10]

At the advice of the old philosopher, the Black Girl and the Irishman are married, have children, and lead useful lives. Eventually the Black Girl comes to see "how funny it was that an unsettled girl should set off to pay God a visit, thinking herself the center of the universe," and as she grows older she goes back to such questions with a strengthened mind that "had taken her far beyond the stage at which there is any fun in smashing idols with knobkerries."[11]

Apparently sometime before April, when the proofs were to be ready, Shaw had dropped some hints to Laurentia about the story, and these were enough to disturb her. So on 14 April 1932 he began a long letter to her: "Your letter has given me a terrible fright. The story is absolutely blasphemous, as it goes beyond all the churches and all the gods. I forgot all about you, or I should never have dared." He then proceeded to outline the story fully and candidly. The final episode of the Black Girl and the Irishman had not yet been added, but he mentioned that he was considering it.

> Perhaps I should not disturb the peace of Stanbrook with my turbulent spirit; but as I want you to go on praying for me I must in common honesty let you know what you are praying for. I have a vision of a novice innocently praying for that good man Bernard Shaw, and a scandalized Deity exclaiming "What! that old reprobate who lives at Whitehall Court, for whom purgatory is too good. Don't dare mention him in my presence."

Then he changes the subject to her election, which, he feels can only be a nominal change "for you would boss the establishment if you were only the scullery maid." But he adds a postscript:

> Shall I send you the story or not? It is very irreverent and iconoclastic, but I don't think *you* will think it fundamentally irreligious. [*IGT*, 254–56]

She demanded to see the book, and dutifully Shaw sent her the bundle of first proofs. On the fly-leaf he wrote:

> An Inspiration
> Which came in response to the prayers of the nuns
> of Stanbrook Abbey
> and
> in particular
> to the prayers of his dear Sister Laurentia
> for
> Bernard Shaw
>
> [*IGT*, 257]

Apparently her response (though not extant) was prompt and outraged. And apparently, too, Shaw sent her, before the end of the month, a somewhat mollifying reply, enclosing a little two-page play. The scene is in God's office in heaven when an argument is taking place between God, Gabriel, and the Recording Angel. God is obviously disturbed at the prayers coming up from Stanbrook Abbey about that fellow Shaw. God has given Shaw a job to do, and the Abbess won't let him do it. The scene closes in on the Almighty's thunderous anger.

Whether it was the implication in the playlet that she was to have the last word, or whether she concluded from Shaw's missing letter that Brother Bernard would do nothing to hurt her, she seems to have mistaken Shaw's cajolery for surrender. In any case, on 3 May she writes:

> You have made me happy again by your nice little play, and I thank you from my heart for listening to me. I have read most of the book and I agree with many of your ideas, but if you had published it I could never have forgiven you. However you are going to be good and I feel light and springy again & proud of my dear Brother Bernard. You shall have more prayers by way of reward! Perhaps you will be naughty enough to set more value on my very deep gratitude. ... I simply cannot find words to thank you for your answer to my letter, but you *know* how grateful I am.

It is impossible to believe that Shaw intentionally led her to believe that he would halt the publication of *The Black Girl*. What reply he made to her letter of 3 May we do not know. But the book, with the delightful engravings of John Farleigh, appeared as a Christmas book in early December, and required five printings totaling 200,000 copies before the end of the year. It may be that there were those who merely glanced at the title and the illustrations and were later embarrassed at the Christmas gift they had chosen for a devout relative or acquaintance!

The Abbess reacted vigorously and uncompromisingly. We do not have her letter of response, but her Benedictine sisters assure us that to the end of her life she could hardly bring herself to mention the book. The scene with Jesus, the image-maker, and Mohammed she considered blasphemy of the most painful kind. She had held both GBS and Cockerell in Christian fellowship, and though she had probably abandoned hope of converting Sir Sydney, she actually harbored a dream of bringing Shaw finally into the Church. Now she could only say that unless Shaw withdrew his hateful work from circulation he would no longer be welcome at the Abbey. Nevertheless, in December, Shaw sent her a copy of the published book inscribed:

> Dear Sister Laurentia,
>
> This black girl has broken out in spite of everything. I was afraid to present myself at Stanbrook in September.
>
> Forgive me
> G. Bernard Shaw
> December 14, 1932 [*IGT*, 260]

Two days later he was aboard the *Empress of Britain* to begin a tour around the world. But the religious argument pursued him. The Abbess's outraged letter caught up with him in Siam. On shipboard he made several unsuccessful attempts to compose a reply. Finally, when he thought he had done the best he could, the notebook in which the letter was drafted disappeared from his deck chair, apparently acquired by "some devoutly Shavian thief." He decided to let the matter rest until he was back at home the following summer. He recounted this misadventure to Laurentia in his letter of 29 June, and tried to reassure her that whoever stole the notebook would not be able to read his shorthand, and, in any case, would not dare to publish it.

> I will not try to reproduce the letter: the moment has passed for that. Besides, I am afraid of upsetting your faith, which is still entangled in those old stories which unluckily got scribbled up on the Rock of Ages before you landed there. So I must go delicately with you, though you need have no such tenderness with me; for you can knock all the story books in the world into a cocked hat without shaking an iota of *my* faith.
>
> Now that I think of it, it was a venial sin to write me such a cruel letter, and I think you ought to impose on yourself the penance of reading *The Black Girl* once a month for a year. I have a sneaking hope that it might not seem so very wicked the tenth or eleventh time as you thought it at first. You must forgive its superficial levity. Why should the devil have all the fun as well as all the good tunes? [*IGT*, 261]

But the old charms no longer worked. "The fact is our points of view are so different, that talking, or writing, round the subject can be of little use," she wrote him. If he wishes to get back into her good grace he must withdraw *The Black Girl* from circulation, and make a public act of reparation for the dishonor the book has done to Almighty God.

> I still have such faith in your greatness of mind & heart, as to believe you capable of such an act, and to ask of you what I should not dream of proposing to a small mind. Do suppress the book & retract its blasphemies, and so undo the mischief it has wrought. I ask you this first & foremost in the interest of your own soul. I have made myself in some sense responsible for that soul of yours, and I hate to see you dishonour it. . . . If you had written against my father or mother, you would not expect to be forgiven or received with any favour until you had made amends. Let me implore you to do this one thing & withdraw the book, even if you cannot find in your heart to imitate St. Augustine & so many other great minds who have given their retractions to the world.[12]

Here, perhaps, the argument should have rested. But Shaw was not able to let it go. Within ten days he was replying:

> You are the most unreasonable woman I ever knew. You want me to go out and collect 100,000 sold copies of *The Black Girl,* which have all been read and the mischief, if any, done; and then you want me to announce publicly that my idea of God Almighty is the antivegetarian deity who, after trying to exterminate the human race by drowning it, was coaxed out of finishing the job by a gorgeous smell of roast meat. Laurentia: has it never occurred to you that I might possibly have a more exalted notion of divinity, and that I dont as a matter of fact believe that Noah's diety ever existed or ever could exist?

And he insisted on the right to his own brand of mysticism:

> . . . what happened was that when my wife was ill in Africa God came to me and said "These women in Worcester plague me night and day with their prayers for you. What are you good for, anyhow?" So I said I could write a bit but was good for nothing else. God said then "Take your pen and write what I shall put into your silly head." When I had done so, I told you about it, thinking that you would be pleased, as it was the answer to your prayers. But you were not pleased at all, and peremptorily forbade me to publish it. So I went to God and said "The Abbess is displeased." And God said "I am God; and I will not be trampled on by any Abbess that ever walked. Go and do as I have ordered you" . . . "Well" I said "I suppose I must publish the book if you are determined that I shall; but it will get me into trouble with the Abbess; for she is an obstinate unreasonable woman who will never let me take her out in my car; and there is no use your going to have a talk with her; for you might as well talk to the wall unless you let her have everything her own way just as they taught it to her when she was a child." So I leave you to settle it with God and his Son as best you can; but you must go on praying for me, however surprising the result may be.
>
> Your incorrigible
> G. Bernard Shaw [*IGT,* 262–63]

But Brother Bernard could not be forgiven. And so for a year or so the relationship languished.

Now what follows sounds like impossible melodrama, and since one of the actors was a skillful creator of melodrama, how can we ever be sure that the incident was not created? Or that the Abbess, taking a page from Shaw's own book, decided to play out her part of the scene as well? In Molnar's *The Guardsman* the husband is never quite sure — nor is the audience — whether his wife has recognized him when he masqueraded as her seducer, and so he can never know whether she has failed the test of constancy which he has so elaboratory arranged.

In September of 1934 Dame Laurentia was celebrating the fiftieth anniversary of her vows. A little folder went out through the mail to announce the jubilee. On the inner page was inscribed:

IN MEMORY OF SEPT. 6
1884–1934
DAME LAURENTIA McLACHLAN
ABBESS OF STANBROOK [*IGT*, 265]

Now either Shaw honestly misread the card, or his Irish sense of anti-climax overcame him and he *decided* to misread it. In any case it was nearly a month later till he dispatched "To the Ladies of Stanbrook Abbey:"

> Through some mislaying of my letters I have only just received the news of the death of Dame Laurentia McLachlan. . . . I never passed through Stanbrook without a really heartfelt pang because I might not call and see her as of old. . . . When my wife was lying dangerously ill in Africa through an accident I wrote a little book which, to my grief, shocked Dame Laurentia so deeply that I did not dare to show my face at the Abbey until I was forgiven. She has, I am sure, forgiven me now; but I wish she could tell me so. In the outside world from which you have escaped it is necessary to shock people violently to make them think seriously about religion; and my ways were too rough. But that was how I was inspired. [*IGT*, 265–66]

The very next day, Laurentia wrote, "As you see, I am not dead," and explained the meaning of the announcement.

> At such a time my mind recalled old friends, you among them. Madame de Navarro said you proposed writing to me, but no letter came, so I took the first step. When next you are in the neighborhood you must come & see me again, and I will tell you some of the wonderful ways in which my nuns have honoured & delighted me on my Jubilee Day.

She is still praying for him, and hopes that her prayers "will have nothing but good results in future." But she signs herself formally: "Sr. Laurentia McLachlan."[13]

Shaw promptly wrote to Cockerell to confess his "super-Howler" and responded to the Abbess:

> Laurentia! Alive!!
> Well!!!!!
> Is this a way to trifle with a man's most sacred feelings?
> I cannot express myself. I renounce all the beliefs I have left. I thought you were in heaven, happy and blessed. And you were only laughing at me!
> It is your revenge for that Black Girl.
> Oh Laurentia Laurentia, Laurentia, how *could* you.
> I weep tears of blood.
>
> Poor Brother Bernard [*IGT*, 266]

And so a reconciliation of sorts was made. Shaw was allowed once more to visit Stanbrook, and the correspondence resumed. But the breach was never wholly healed. In January of 1935 he sent her, at her request, the proofs of *The Simpleton of the Unexpected Isles*, with the warning on the flyleaf that "if you can get over the first shock of its profanity you may find some tiny spark of divinity in it. You may ask why I write such things. I dont know. I have to. The devil has me by one hand and the Blessed Virgin by the other" (*IGT*, 268).

Laurentia had little trouble finding the divinity, but she was not about to excuse the profanity. Once again she asked him not to let the work be published. Strangely it was not the play's eugenics experiment or its parody of the Judgment Day to which she objected so much as the insistence of the Island Priest that Christianity was as polytheistic as any Oriental religion.

> THE PRIEST. God has many names.
> THE LADY TOURIST. Not with us, you know.
> THE PRIEST. Yes: Even with you. The Father, the Son, the Spirit, the Immaculate Mother. . . . [VI, 777]

"Whatever happens," Laurentia wrote, "you absolutely must omit the allusion to the Immaculate Mother. . . . You know, as well as I do, that we do not worship her as God." Beyond that the play might be harmless to "strong minds but it is almost certainly dangerous for weak ones." He has given the devil enough innings, she thinks, and it is time to let Our Lady take him by *both* hands. And she reminds him of his age. "However you may parody the Day of Judgment, you know the particular one for each of us can't be very far off and, my Dear Brother Bernard, you will not be able to plead ignorance as the excuse of the evil that your books may do" (*IGT*, 268–69). Shaw was then approaching seventy-nine. She was sixty-nine.

But the "Immaculate Mother" remained in the published play, and there was no evidence in Shaw's reply that he had any concern about meeting his Maker. Instead he challenged the legitimacy of her Catholicism. "Why in the name of all the saints does she fly out at me when I devoutly insist that the Godhead must contain the Mother as well as the Father?" Does he say Hail Mary? Of course he does, but he says it in his own natural and sincere way, whenever She turns up in any of the foreign shrines or temples that he visits: "Hallo, Mary!"

> When I write a play like *The Simpleton* and have to deal with divinity in it She jogs my elbow at the right moment and whispers "Now Brother *B.* dont forget *me*." And I dont. . . . When you are old, as I am, these things will clear up and become real to you. . . . It is in these temples [of the Jains] that you escape from the frightful parochiality of our little sects of Protes-

> tants and Catholics, and recognize the idea of God everywhere, and understand how the people who struggled hardest to establish the unity of God made the greatest number of fantastically different images of it, producing on us the effect of a crude polytheism. [*IGT*, 271–72]

This time there were no threats that he would be banished from Stanbrook or that the letters would cease. But as they both became older and as health became more precarious the visits dwindled off, and the correspondence became a birthday exchange or a "thank you" for the latest book. He visited her at the end of August 1935, and found her "shining in all your old radiance before the cloud of illness came upon you," and he confesses that he "felt ever so much better for your blessing." But he is careful never to leave too great an impression of piety: "There are some people, who, like Judas Iscariot, have to be damned as a matter of heavenly business; . . . but if I try to sneak into paradise behind you they will be too glad to see you to notice me . . ." (*IGT*, 273). Of *The Millionairess* she had no complaints. It "entertained me immensely."[14] From her letter of 20 August 1938 it is apparent that they have both been ill. She wants an inscribed copy of the new play (probably *Geneva*). There are hints that Stanbrook needs money. I can find no record that Shaw made financial contributions to the Abbey, though it is possible that he may have.

Charlotte's failing health finally brought an end to the Shaws' travels. She died in September of 1943 at the age of eighty-six. Laurentia's note of 29 August 1944 finds her thanking him for "the admirable portrait of Mrs. Shaw, whom I am not likely to forget." A few days later Shaw sends her a note on her Diamond Jubilee, written on the flyleaf of his new work, *Everybody's Political What's What?* "The saint who called me to the religious life when I was eighteen was Shelley. But you have lived the religious life: I have only talked and written about it. . . . You would still know me if you met me. I wish you could. I count my days at Stanbrook among my happiest."

In her reply (19 September) she reaffirms that

> Sixty years of enclosed life leaves me happier than ever for having chosen this path — though as Tagore sang: "We cannot choose the best. The Best chooses us." If people knew the freedom of an enclosed nun! I believe you can understand it better than most people & I only wish we shared the faith that is its foundation.

During the week of his ninetieth birthday (26 July 1946) Shaw received over a hundred congratulations each day. Not many received a personal reply, but to Laurentia he wrote,

> Saving your reverence, I do not give a damn for congratulations. But

> prayers touch me and help me. It is good for me to be touched. Stanbrook prayers must have some special charm; for I never forget them. [*IGT*, 274]

At ninety-two he wrote her a fairly long letter recounting the visit of Gene Tunney, and had obvious pleasure in transmitting the story of how the former heavyweight champion of the world went back to his Catholic faith and prayed for help when his wife was stricken with double appendicitis on an island in the Adriatic, and how "Next morning very early there landed in the island the most skilful surgeon in Germany, the discoverer of double appendices. Before ten o'clock Mrs. Tunney was out of danger and is now the healthy mother of four children" (*IGT*, 275).

To the outside world Shaw's longevity seemed evidence that he was indeed one of the supermen of whom he had written in *Back to Methuselah*, and his birthdays became more and more of an event to be noted and celebrated. His house was surrounded by photographers and reporters, and he became the protesting recipient of cakes, letters, phone calls, and gifts, including "piles of medals of the Blessed Virgin, with instructions that if I say a novena she will give me any help I ask from her; and I have to reply that we are in this wicked world to help her and not to beg from her" (*IGT*, 276). Nevertheless Laurentia added one more medal to the pile. On the 8th of August 1950, just three months before his death, she sent him a little bag of lavender from the harvest that had just come in (and which the nuns bagged for sale). "There is a little medal of Our Lady buried in the lavender, but you will not object to that."

Even the nonagenarian epistles are newsy, friendly, and occasionally disputatious, and he continued to send her copies of everything he published. If the mistaken announcement of 1934 was a ruse on his part, or on the part of both, it was played out to the end. Dame Laurentia survived him by three years. She died at age eighty-seven.

The relationship between Shaw and Laurentia remained almost unknown until Blanche Patch published her memoirs the year after Shaw's death. Shaw's liaisons with women, after his philandering years in 1880s, were mostly on paper. The passionate love-letters to Ellen Terry in the 1890s, too, are full of cajoling and scolding and lecturing. The affair with Stella Campbell may have had more substance, but that correspondence is chiefly an account of carefully contrived pursuits and escapes. In the course of both of these exchanges the female respondents married other men — to Shaw's expressed relief!

The correspondence with Laurentia may have been something of a supersession for those stormy days of his middle years. These letters are not as obviously love-letters, but, in a sense, what else are they? Throughout the religious argument there is always apparent the surge of affection and the need for her approval. Though he would not yield to her pleas to alter his published work, he changed his entire theological vocabulary to remain in her good graces. In none of his religious writings or speeches does he come as close to the orthodox vocabulary of the Church as he does for her sake. His attitude toward prayer, as expressed in these letters, may have been sincere, but it would have surprised most of his associates, and is revealed nowhere else in quite the same way.

Shaw, the great feminist, was always somewhat afraid of *real* women — beginning with his own mother. He needed the feminine response to complement his own masculine assertiveness; but he feared the softness and sentimentality to which he knew he was vulnerable, but which were counter to the Shaw message. In this regard Dame Laurentia was an ideal respondent. Even more than Ellen Terry, she was inaccessible.

And for her part, once she had admitted this beguiling Irishman into her confidence, she found that she could not do without him. Her relief when Shaw resumed the correspondence at her "death" notice is palpable. Though she extols the enclosed life, this breath of heresy from the freethinking world nourished her and brightened her life. And her natural warmth responded to the blandishment that was safely on the other side of the grille.

Certainly her prayers and her concern for Shaw's immortal soul were genuine. But even towards the end she must sometimes have felt uneasy when she recalled what he had told her at the height of *The Black Girl* controversy: "You must go on praying for me, *however surprising the result may be.*"

Envoy

Reasonable Mystic, Laughing Prophet

The survival of the Life Force as a hypothesis does not depend solely on the survival of Shaw. From Bergson through Teilhard de Chardin to Arthur Koestler the idea has had its champions, though not always under that term. It is not in current favor in the scientific community. We are in an era when nothing is accepted as scientific unless it can be measured or counted. And since the Life Force, if it exists, must always lie back of what can be counted or measured, it is not likely to find its way into the biological or anthropological treatises of our times. Still, the notion that human consciousness is nothing but a happy accident in a mindless universe repels not only our natural vanity, but also what we are pleased to call our common sense. As Shaw put it:

> The professional biologists tell us nothing of all this. It would take them out of the realm of logic into that of magic and miracle, in which they would lose their reputation for omniscience and infallibility. But magic and miracle, as far as they are not flat lies, are not divorced from facts and consequently from science: they are facts: as yet unaccounted for, but none the less facts. As such they raise problems; and genuine scientists must face them.... [VII, 310]

So it is probable that the Life Force, by whatever name, will continue to inhabit our thinking whether mechanistic science approves or not. And it is probable that Bernard Shaw will continue to be its leading prophet. I cannot guess how long people will continue to read his letters and treatises. Certainly his comprehension of the problems of government may seem to us inadequate or even (in the light of his fulsome admiration of Josef Stalin) naive. But now, as we near the end of the century he helped to usher in, it should no longer be necessary to enroll either as a disparager or a

Shavolator. It ought to be possible to recognize his profound contributions to the twentieth-century mind without having to accept him as an infallible guru on every subject on which he chose to pontificate. We ought to be prepared, in other words, to recognize Shaw's shortcomings as Samuel Johnson was prepared to recognize Shakespeare's: "without envious malignity or superstitious veneration." The fact that he had foibles of his own does not invalidate his clear-headed perception of the foibles of the rest of humanity.

There can be no doubt that some of his plays are enduring achievements, and that they carry the message of the Life Force more persuasively than his more patent sermons on the subject. Of course, he himself has warned us that works of art do survive beyond the beliefs that inspired them:

> All the assertions get disproved sooner or later, and so we find the world full of a magnificent débris of artistic fossils, with the matter-of-fact credibility gone clean out of them, but the form still splendid. [II, 527–28]

The best of Shaw's plays are not yet "artistic fossils." Of his more than fifty it is possible that only a half dozen or so are likely to remain in the standard repertory (though I believe many more of them will continue to be read and studied). His influence on the latter half of the twentieth century should not be underestimated. John Gassner in his article on "Bernard Shaw and the Making of the Modern Mind"[1] reminds us that almost every so-called "advanced" idea of our times was explicated by Shaw. (But at this point we hear Shaw's chuckle in the wings, as we recall Jack Tanner's reply to Roebuck Ramsden's claim that "I was an advanced man before you were born": "I knew it was a long time ago" (II, 547).)

Even at the risk of this classic put-down, can it be denied that there is something perennially challenging and refreshing about this man and his desire to save us all from poverty, respectability, hypocrisy, organized education, meat, the crime of imprisonment, aphonetic spelling, the worship of Shakespeare, and the thousand other ills that made this world hell for Shaw and Peter Keegan? The world has had its share of prophets, a number of whom have preached quite similar doctrines. With what had the Life Force endowed this one that made him so different, so irreplaceable?

Mostly, I think, it was his laughter. A laughing prophet of real stature was something new in the world. Tolstoy made the mistake of thinking the laughter sprang from levity. But the laughter of Shaw was neither that of frivolity nor of Byronic bitterness nor even very often of plain ribaldry (though occasionally it was that). For the most part it was the laughter of immense good health and irrepressible good feeling. It was the mark of determination not to get too involved in the world, a sort of emotional declaration of independence — which, in his later years, he sometimes failed to maintain. The laughter was part of his religion, too. The theatre, indeed, was a place "where two or three are gathered together," and he claimed an "apostolic succession" from Aeschylus. The "younger institution," the Christian Church, was, he insisted, "founded gaily with a pun"—referring apparently to the account in Matthew where Jesus says, "Thou art Peter, and on this rock I will found my church," "Peter" and "rock" being the same Greek word. (There is a similar account in John, where the Gospel writer uses the term *Cephas* for "rock.") It is not really clear in either of these scriptures whether Jesus was making an apt use of Simon Peter's surname, or *giving* him the name "Rock." In any case Shaw is probably the first to find humor in these passages! But unfortunately the Church has degenerated into a place where you may not laugh,

> ... and so it is giving way to that older and greater Church to which I belong: the Church where the oftener you laugh the better, because by laughter only can you destroy evil without malice, and affirm good fellowship without mawkishness.[2]

Much of the laughter came out of his gift for paradox. But it annoyed him to have people think that he invented paradoxes merely for laughter's sake. He simply could not help being conscious of the incongruities in our topsy-turvy world. As he had Peter Keegan say, "My way of joking is to tell the truth. It's the funniest joke in the world" (II, 930). And, of course, his own nature was full of paradoxes, not the least puzzling of which was his image as a rationalist. Not that he ever claimed to be one. It is our own mistaken assumption that so much brainpower, so beautifully balanced a sentence, so devastating a dialectic must of necessity speak with the voice of reason. In 1912 when the *New York Times* asked him "to define the principles that govern the dramatist in his selection of themes and methods of treatment," he replied, "Who told you, gentlemen, that dramatists are governed by principles? ... I am not governed by principles; I am inspired, how or why I cannot explain, because I do not know; but inspiration it must be; for it comes to me without any reference to my own ends or interest."[3]

A strange mystic indeed! Without the need either for a halo or for overweening humility, he nevertheless found a power that propelled his characters across the stage, and used them for its purposes almost as the Greek gods would have used them.

It is just possible that the reasonable mysticism of this laughing prophet, proceeding from the Life Force (as he claimed) and with all its paradoxes intact, may be one of the more reliable guidelines for the fearful times that lie ahead.

Notes

Complete publication information, if not given in the notes, is available in the bibliography which follows.

PROLOGUE: SHAW VS. SHAW VS. SHAW

1. All quotations from Shaw's plays and their prefaces are from *Collected Plays and their Prefaces*, ed. Dan Laurence; volume and page numbers will be cited in the text at the end of each quotation.
2. "On Going to Church," in Stanley Weintraub, ed., *The Savoy, Nineties Experiment*, p. 12.
3. B.M. 50720.
4. Archibald Henderson, *George Bernard Shaw, Man of the Century*, p. 29.
5. See also Arthur Nethercot, *Men and Supermen*, pp. 91, 310, n. 16.
6. The entire poem is quoted in my article, "An Early GBS Love Poem," *Shaw Review*, X:2 (May 1967), 70–72.
7. Warren S. Smith, *The London Heretics, 1870–1914*, pp. 223–35.
8. Raymond S. Nelson, "Shaw's Keegan," *Shaw Review*, XIII:3 (May 1970), 92–95.

CHAPTER 1: THE TEMPTATIONS OF JUDAS ISCARIOT

1. C.E.M. Joad, *Shaw*.
2. Eric Bentley, *Bernard Shaw*.
3. *The Black Girl in Search of God and Some Lesser Tales*, p. 7.
4. *Sixteen Self Sketches*, p. 74.
5. Shaw's Preface to Richard Albert Wilson's *The Miraculous Birth of Language* (New York: Philosophical Library, 1948), p. 19.
6. "The Infancy of God," in *Shaw on Religion*, pp. 132–33.
7. Preface to *Immaturity*, p. xliii.
8. Ibid., pp. xviii, xix.
9. The entire letter is quoted in Henderson, *Man of the Century*, pp. 47–48.
10. Notably unconvinced was Ben Rosset. See his *Shaw of Dublin*.
11. This fragment was also published separately under the title, *Passion Play, A Dramatic Fragment, 1878*, edited by Jerald E. Bringle (Iowa City: Windhover Press, 1971). For a penetrating analysis of this early work and the influence of Shelley's poëtry on Shaw, see Charles A. Berst's "In the Beginning: The Poetic Genesis of Shaw's God," in *SHAW I*, pp. 5–41.
12. Reference is to a typed letter, signed, to J. A. Hughes, 5 January 1938; Bucknell

University Library, Lewisburg, Pennsylvania.
13. Preface, *Immaturity,* p. xxxiv.
14. *Shaw, An Autobiography,* Vol. I, p. 299.
15. *Platform and Pulpit,* p. 131.
16. Benedictines of Stanbrook, *In a Great Tradition,* p. 241.
17. The article was written for T. P. O'Connor's *Star* but not published. Dan H. Laurence printed it in the *Flying Dutchman,* Hofstra College's student newspaper, 23 November 1957, p. 3.
18. Letter from Shaw to F. H. Evans, 27 August 1895, in Shaw's *Collected Letters, 1874–1897* (Vol. I), p. 551. Volume II contains letters from 1898 through 1910. Hereafter referred to as *Letters I* and *Letters II.*
19. Weintraub, *The Savoy,* pp. 3–15.

CHAPTER 2: EVOLUTION OF THE SUPERMAN

1. Christopher St. John, ed., *Ellen Terry and Bernard Shaw, a Correspondence,* letter of 4 August 1899, p. 246.
2. Charles Carpenter, *Bernard Shaw and the Art of Destroying Ideals,* p. 216.
3. Ibid., p. 215:
Widowers' Houses — comic propaganda
The Philanderer — farcical propaganda
Mrs Warren's Profession — melodramatic propaganda
Arms and the Man — romantic comedy
Candida — sentimental comedy
The Man of Destiny — heroic drama (of sorts)
You Never Can Tell — comedy of manners
The Devil's Disciple — comic melodrama
Caesar and Cleopatra — modern heroic drama
Captain Brassbound's Conversion — melodramatic comedy
4. Maurice Valency, *The Cart and the Trumpet,* pp. 204–205.
5. F. H. Bradley, *Ethical Studies* (1876); *Appearance and Reality* (1893).
6. E. Strauss, *Bernard Shaw: Art and Socialism,* p. 41.
7. St. John Ervine, *Bernard Shaw, His Life, Work, and Friends,* pp. 368 ff.
8. St. John, *Terry-Shaw Correspondence,* p. 126.
9. Ibid., p. 185.
10. Charles Berst, *Bernard Shaw and the Art of the Drama,* pp. 145 ff.
11. See Daniel Dervin, *Bernard Shaw, a Psychological Study,* pp. 105, 245n.
12. St. John, *Terry-Shaw Correspondence,* p. 177.
13. *Letters II,* p. 426.
14. *Shaw Autobiography, II,* p. 151. Quoted from *The Fortnightly Review,* April, 1926.
15. Alfred Turco, Jr., *Shaw's Moral Vision* pp. 145, 155.
16. Foreword to the Popular Edition of *Man and Superman,* 1911.
17. *Letters II,* p. 873. Letter to Julie Moore.
18. Ibid., 902.
19. Valency, *Cart and Trumpet,* p. 201.
20. Norman Holland, *The Dynamics of Literary Response* (New York: Oxford University Press, 1968), p. 24. See also A. M. Gibbs, "Comedy and Philosophy in *Man and Superman,*" *Modern Drama,* XIX:2 (1976), 161–175; and J. S. Wisenthal, "The Cosmology of *Man and Superman,*" *Modern Drama,* XIV:3 (1971), 298; and Berst, *Art of the Drama,* pp. 96–7.
21. Bentley, *Bernard Shaw,* p. 154.
22. Holland, *Literary Response,* p. 247.
23. Samuel Butler, *Luck or Cunning?,* p. 234.
24. Gibbs, "Comedy and Philosophy."
25. *Letters II,* p. 528.
26. Louis Crompton, *Shaw the Dramatist,* p. 76.
27. Margery Morgan, *The Shavian Playground,* p. 102.

28. Daniel J. Leary, "Shaw's Use of Stylized Characters and Speech in *Man and Superman*," *Modern Drama*, V:4 (1963), 477–90.

29. Quoted from *The New York Times* of 6 January 1929, in Henderson, *Man of the* Century, p. 727.

30. Quoted in Smith, *London Heretics*, p. 19.

31. Stanley Weintraub, "The Embryo Playwright in Shaw's Early Novels," *Studies in Literature and Language*, I (1959), 341.

32. "How Frank Ought To Have Done It" in *Sixteen Self Sketches*, p. 129.

33. *Shaw on Religion*, pp. 181–83.

34. Daniel J. Leary explores these Freudian implications in "Don Juan, Freud, and Shaw in Hell: A Freudian Reading of *Man and Superman*," *The Shaw Review*, XXII:2 (May 1979), 58–78.

35. Richard Ohmann, *Shaw, the Style and the Man*, p. 106.

36. Crompton, *Shaw the Dramatist*, p. 104.

37. Turco, *Shaw's Moral Vision*, p. 169.

38. Berst, *Shaw and Drama*, pp. 109, 122.

39. *Platform and Pulpit*, p. 174.

40. Quoted by Valency, *Cart and Trumpet*, p. 218.

41. Carl Henry Mills, "Shaw's Debt to Lester Ward," *The Shaw Review*, XIV:1 (1971), 3–17.

42. John Gassner, "Bernard Shaw and the Making of the Modern Mind," *College English*, XXIII (1962), 517–525. Reprinted in the Norton Critical Edition of *Bernard Shaw's Plays*, edited by W. S. Smith (New York, 1970).

43. Berst, *Shaw and Drama*, p. 104.

44. Morgan, *Shavian Playground*, p. 118.

45. Quoted in Henderson, *Man of the Century*, pp. 582–83.

46. Berst, *Shaw and Drama*, p. 103.

47. Morgan, *Shavian Playground*, p. 118.

48. Turco, *Shaw's Moral Vision*, p. 146.

49. Nethercot, *Men and Superman*, p. 279.

50. From the note accompanying the program of the Royal Court Theatre, London, 4 June 1907.

51. T. F. Evans, ed., *Shaw, the Critical Heritage*, p. 97.

52. Ibid., pp. 110–114.

53. Ibid., p. 117.

54. Allan Rodway, *English Comedy*, (Berkeley: University of California Press, 1975), pp. 237–43.

55. Berst, *Shaw and Drama*, p. 233.

56. Ibid., p. 96.

57. Valency, *Cart and Trumpet*, p. 202.

58. Ervine, *Shaw: Life, Work, Friends*, p. 368.

59. Quoted in Henderson, *Man of the Century*, p. 766.

60. See Sidney Albert's article, "'In More Ways Than One': Major Barbara's Debt to Gilbert Murray," *Educational Theatre Journal*, XX (1968), 123–40.

CHAPTER 3: PREACHER OF A NEW RELIGION

1. Ernst Heinrich Haeckel (1834–1919) was a German zoologist, doctrinaire evolutionist, and mechanist, with considerable influence in the Secularist movement.

2. "Modern Religion," a speech delivered at the New Reform Club, London, 21 March 1912; *Religious Speeches* p. 40. Shaw actually made two speeches to the Secularists, one to the Camberwell Branch on 26 April 1891, and one at the Hall of Science on 4 August 1892. In his 1912 recollection he may be confusing or amalgamating these two. It would appear from a letter written to E. C. Chapman on 29 July 1891 that it is the earlier address Shaw is describing rather than the Hall of Science one. *Letters I*, pp. 300–03.

3. *Religious Speeches*, p. xix. The reporter was Albert Dawson of the *Christian Commonwealth*.

4. Ibid., pp. xix, xx. The reporter was R. B. Suthers.
5. Ibid., p. xx.
6. The manuscript is in the British National Library, marked in Shaw's hand, "Discards from the Fabian Lectures On Ibsen and Darwin when publishing them as The Quintessence of Ibsenism and the Methuselah Preface." Ps 80512/59 ADD. 50661. The references that follow are from that manuscript. (See also Shaw's comment in the Preface to *Back to Methuselah*, V, 260.)
7. Ibid., p. 50.
8. Ibid., pp. 49–50.
9. Ibid., p. 80.
10. Ibid., p. 97.
11. Ibid., p. 49.
12. Ibid., p. 56.
13. Ibid., p. 66.
14. Ibid., pp. 93–94.
15. But Charles A. Berst thinks that *Major Critical Essays* (1932) constitutes the "Gospel of Shawianity," particularly the second of these essays, "The Perfect Wagnerite," written in 1898. See Berst's "Poetic Genesis of Shaw's God" in *Shaw I*, pp. 5–39.
16. Henderson, *Man of the Century*, p. 31.
17. Annas and Caiaphas were the high priests who turned Jesus over to Pilate. John 18:23 ff.
18. *Letters I*, p. 197.
19. See Shaw's comment in the Preface to *Farfetched Fables*, VII, 396.
20. This summary has been adapted from my Introduction to *Religious Speeches*, pp. xxii–xxiii.

CHAPTER 4: INTERLUDE WITH A BISHOP AND A DANCER

1. A full account of Shaw's struggles with the Lord Chamberlain's office, and his efforts to have that office abolished, appears in the Preface to *The Shewing-Up of Blanco Posnet*, III, 675–762.
2. All quotations from *The Times* of London are from the issues of 1 November through 15 November 1913.
3. Shaw is quoting, not quite accurately, from Revelation 22:11.
4. As reported by Henderson in *Table-Talk of G.B.S.*, pp. 128–34.

CHAPTER 5: PACIFISM AND THE QUAKERS

1. *What I Really Wrote About the War*, p. 189.
2. See "Crude Criminology" in *Doctors' Delusions, Crude Criminology, and Sham Education*, p. 192.
3. "Religion and War" in *Religious Speeches*, p.96.
4. "If I Were a Priest," *Atlantic Monthly*, Vol. 185, No. 5 (May 1950), 7.
5. Blanche Patch, *Thirty Years With Bernard Shaw*, p. 227.
6. S. J. Woolf, *Here Am I* (New York: Random House, 1941), p. 181.
7. Stephen Winsten, *Days With Bernard Shaw*, p. 184.
8. Preface to Wilson, *Miraculous Birth of Language*, p. 19.
9. *Letters II*, pp. 121–23.
10. *What I Really Wrote*, p. 188.
11. Ibid., p. 190.
12. Ibid., p. 90.
13. Ibid., p. 189.
14. See the account of Stanley Weintraub in *Journey to Heartbreak*, pp. 252–53.
15. Quoted from *The New York Times*, 7 June 1916, in Weintraub, *Journey to Heartbreak*, p. 161.
16. BBC broadcast, 2 November 1937, in *Religious Speeches*, p. 96.
17. *What I Really Wrote*, p. 314.

18. *Religious Speeches*, p. 95.
19. Hesketh Pearson, *G.B.S. A Full-Length Portrait*, p. 357.

CHAPTER 6: THE STRUGGLE AGAINST CYNICISM

1. Charles Berst convincingly suggests that he was equally influenced by Tolstoy, Strindberg, and Pirandello. Berst, *Shaw and Drama*, p. 253.
2. Aubrey in *Too True To Be Good*.

CHAPTER 7: THE PROBLEM: WHY?

1. Julian Huxley, *Evolution in Action*, p. 35.
2. Huxley, *Evolution in Action*, p. 72.
3. Julian Huxley, *Memories*, p. 252.
4. Julian Huxley, *Heredity, East and West*, pp. 210–11.
5. Ibid., p. 139.
6. Huxley's Introduction to Pierre Teilhard de Chardin's *The Phenomenon of Man*, p. 11.
7. For an account of Teilhard's troubles with his superiors, see Martin Jarrett-Kerr's Preface to the English edition of Nicolas Corte's *Pierre Teilhard de Chardin, His Life and Spirit* (New York: Macmillan, 1960).
8. Julian Huxley, *Essays of a Biologist*, p. 193.
9. Huxley, Preface, *Essays of a Biologist*, p. 11.
10. In his address to the University of Chicago's Centennial Celebration of the publication of *The Origin of Species*, 1959. Julian Huxley, *Essays of a Humanist*, p. 78.
11. Introduction to Teilhard, *Phenomenon of Man*, p. 26.
12. Daniel J. Leary, "The Evolutionary Dialectic of Shaw and Teilhard: A Perennial Philosophy," *Shaw Review*, IX:1 (January 1966), 15–34.
13. Leary, "Shaw and Teilhard," p. 17.
14. Charles Darwin, *The Origin of Species and The Descent of Man*, pp. 160–62.
15. Jacques Monod, *Chance and Necessity*, p. 104.
16. Ibid., pp. 112–113.
17. Arthur Koestler, *The Case of the Midwife Toad*, p. 129.
18. In a paper read to the Association for General and Liberal Studies at The Pennsylvania State University on 10 November 1979, "Reincarnation of Natural Philosophy," Dr. Gibbs estimated the number of possible sequences in a DNA molecule to be $10^{1,000,000}$; whereas the beginnings of life can be placed at a mere 10^{17} seconds ago.

CHAPTER 8: LAMARCK VS. DARWIN

1. J. B. Lamarck, *Zoological Philosophy.* The 92-page Introduction by Hugh Elliot, the translator, dates from 1914 and provides a biography and a very helpful summary of the work.
2. Teilhard, *Phenomenon of Man*, pp. 149–50 n.

CHAPTER 9: THE LIFE FORCE, THE NOOSPHERE AND THE NEW RELIGIONS

1. Pierre Teilhard de Chardin, *Christianity and Evolution*, p. 238. The italics are always Teilhard's unless otherwise noted.
2. Teilhard, *Phenomenon of Man*, p. 182.
3. Pierre Teilhard de Chardin, *The Future of Man*, pp. 47–49.
4. Teilhard, *Phenomenon of Man*, p. 56.
5. Ibid., p. 61.
6. Teilhard, *Future of Man*, p. 229.
7. Ibid., pp. 290–1.
8. See especially the first three of the *Religious Speeches*; and the 1905 address, "Life, Literature, and Political Economy," in *Practical Politics*, pp. 1–9.
9. Bertrand Russell, *A History of Western Philosophy* (New York: Simon and Schuster, 1945), p. 791.

10. Henri Bergson, *Creative Evolution*, p. 58.
11. Ibid., pp. 90 ff.
12. Ibid., p. 97.
13. Teilhard, *Christianity and Evolution*, p. 240.
14. Introduction, Teilhard, *Phenomenon of Man*, p. 19. See also "The Humanist Frame" and other essays in Huxley, *Essays of a Humanist*, passim.
15. *Religious Speeches*, p. 35.

CHAPTER 10: OUR PLACE IN THE PRESENT UNIVERSE

1. Huxley, Preface, *Essays of a Biologist*, p. 11.
2. Ibid., p. 37.
3. Teilhard, *Future of Man*, p. 266.
4. Ibid., p. 310.
5. Ibid., p. 243.
6. Ibid., pp. 145 ff.
7. Russell F. Knutson, ed., *Shaw on Vivisection* (Chicago: Alethea Publications, 1950), p. 23.
8. Teilhard, *Future of Man*, p. 242.
9. See the last three essays in Huxley's *Essays of a Humanist*.
10. Richard E. Leakey and Roger Lewis, *Origins* (New York: E. P. Dutton, 1977), p. 208.
11. Julian Huxley, *Man Stands Alone*, p. 288.
12. St. John, *Terry-Shaw Correspondence*, Preface, p. xxviii.

CHAPTER 11: THE FUTURE OF THE HUMAN RACE

1. For a comprehensive account see "Eugenics in Evolutionary Perspective" in Huxley's *Essays of a Humanist*.
2. Teilhard, *Future of Man*, p. 236, italics mine.
3. Ibid., p. 244.
4. *Parousia* is the Greek theological term for "the coming" or the Advent of Christ.
5. Teilhard, *Future of Man*, pp. 246–47.
6. Bergson, *Creative Evolution*, p. 50, italics mine.
7. Huxley, *Essays of a Biologist*, p. 40.
8. Stephen Jay Gould, *Ever Since Darwin*, p. 61.
9. Huxley, *Essays of a Humanist*, p. 51.
10. "A Glimpse of the Domesticity of Franklyn Barnabas," prefatory comments to a discarded scene.
11. Teilhard, *Phenomenon of Man*, p. 79. The more recent discovery of methanogens, an extremely primitive form of life, would tend, however, to indicate that even the emergence of life was a gradual process that proceeded in very small steps. The above statement of Teilhard also appears to be somewhat in conflict with his own views, developed at greater length elsewhere, that there can be no clear line between life and non-life.
12. Teilhard, *Future of Man*, p. 205.
13. Ibid., p. 213, italics mine.
14. See especially Part II of Arthur Koestler, *The Ghost in the Machine*.
15. As reported by Bryan Silcock in the *Sunday Times*, London, 3 July 1977, based on a report to *Nature* by Mario Coluzzi et al., 28 April 1977, Vol. 266, pp. 832–33. Further support for the hypothesis of "punctuational" evolution is now given by Prof. Steven M. Stanley in his *The New Evolutionary Timetable* (New York: Basic Books, 1982).
16. Huxley, *Essays of a Biologist*, pp. 34–35; *Essays of a Humanist*, p. 37.
17. Teilhard, *Future of Man*, pp. 142–43.
18. Shaw to St. John Ervine, quoted in Ervine, *Shaw's Life, Work and Friends*, p. 490.
19. Teilhard, *Future of Man*, p. 92.
20. Ibid., p. 292.

21. Teilhard, *Phenomenon of Man*, p. 280.
22. Darwin, *Descent of Man*, p. 543.

CHAPTER 12: "AS FAR AS THOUGHT CAN REACH"

1. Huxley, *Essays of a Humanist*, pp. 198 ff.
2. Teilhard, *Future of Man*, p. 181.
3. *Letters II*, p. 902.
4. Teilhard, *Future of Man*, p. 266.
5. Teilhard, *Phenomenon of Man*, p. 283.
6. Teilhard, *Future of Man*, p. 311.
7. Ibid., p. 127.
8. Revelation 1:8.
9. Teilhard, *Phenomenon of Man*, p. 191.

CHAPTER 13: IN SUMMARY

1. Leary, "Shaw and Teilhard," p. 32.
2. Huxley, *Essays of a Biologist*, p. 38.
3. Teilhard, *Phenomenon of Man*, pp. 177–78.

CHAPTER 14: THE WEB OF AMBIGUITY

1. *Major Critical Essays*, p. 135.
2. Frederick P. W. McDowell has given this neglected play a long-overdue critical analysis in "Shaw's Abrasive View of Edwardian Civilization in *Misalliance*," *Shaw Review*, XXIII:2 (May 1980), 63–76.
3. Leon Hugo, *Bernard Shaw, Playwright and Preacher*, p. 196.
4. Crompton, *Shaw the Dramatist*, p. 208.
5. In "What Is My Religious Faith?" *Sixteen Self Sketches*, p. 79.
6. *The Intelligent Woman's Guide*, p. 302.
7. Quoted in Lynn Thorndike, *History of Medieval Europe* (New York: Riverside Press, 1928), p. 639.
8. In W. K. Wimsatt, Jr., ed., *English Stage Comedy* (New York: Columbia University Press, 1954), p. 136.

CHAPTER 15: THE FAILURE OF GOVERNMENTS

1. The earliest of the *Essays in Fabian Socialism* are from the 1880s. The volume in the Standard Edition is dated 1932. *Everybody's Political What's What?* was first published in 1944.
2. Henderson, *Man of the Century*, p. 319.
3. Raymond Mander and Joe Michenson, *Theatrical Companion to Shaw*, p. 262.
4. See "The Transition to Social Democracy," *Essays in Fabian Socialism*, p. 47.
5. "The Simple Truth about Socialism" in *The Road to Equality*, pp. 184–85.
6. See for instance his October 1929 broadcast which is included in the Preface to *The Apple Cart*, VI, 256–73.
7. *Intelligent Woman's Guide*, p. 337.
8. See his speech at the Metropolitan Opera House, New York, as reported in *The New York Times* 12 April 1933, p. 14, later published under the title of *The Political Madhouse in America and Nearer Home*.
9. *Everybody's Political What's What?*, p. 338.
10. Ibid., p. 340.
11. Ibid., p. 339.
12. Ibid., pp. 339–40.
13. *Intelligent Woman's Guide*, p. 379.
14. *Everybody's Political What's What?*, pp. 309–10.
15. Ibid., pp. 336 ff.
16. Ibid., p. 337.

CHAPTER 16: POSTLUDE WITH A NUN AND *THE BLACK GIRL*

1. The principal author was Dame Felicitas Corrigan, O.S.B.
2. See chap. 5, note 5, supra.
3. Benedictines, *In a Great Tradition* (*IGT*), p. 255. All the letters from Shaw are from this volume. Page references hereafter are incorporated into the text.
4. The Standard Edition, p. xxiv.
5. Letter dated 1 March 1931. All the letters from Dame Laurentia McLachlan are a.l.s. and are from the Manuscript Collection of the British Museum (Add. Mss. 50543).
6. Letter dated 18–21 Apr. 1931. In an earlier letter, erroneously dated "27 iii 30" instead of "27 iii 31," she thanks Shaw for the olive leaves and wonders whether she could bear to visit the Holy Land "under modern conditions."

This letter also indicates the degree of intimacy that had developed between the correspondents. In his September letter Shaw had shared with her his concern about the possible publication of the Ellen Terry-Bernard Shaw Correspondence, complaining that the letters had fallen into the hands of an American speculator who was under the impression that he had purchased the rights to publication. Laurentia replies:

> [Miss] Christopher [St. John] tells me that *the* letters are coming out. From your letter in September I gathered that you had no choice left in the matter. At any rate their appearance brings great contentment to Bedford St., and your generosity a certain ease for which I am very grateful. I hope Christopher will now be free to use her talent to some purpose.

7. 12 Oct. 1931.
8. 30 Oct. 1931.
9. *The Black Girl*, p. 54.
10. Ibid., p. 60.
11. Ibid., p. 72.
12. 13 July 1933.
13. 4 Oct. 1934.
14. 9 Sept. 1935.

ENVOY: REASONABLE MYSTIC, LAUGHING PROPHET

1. See chap. 2, note 43, supra.
2. *Our Theatres in the Nineties*, Vol. I, p. vi.
3. *New York Times*, 2 June 1912; reprinted in Barrett H. Clark, ed., *European Theories of the Drama* (New York: D. Appleton-Century Co., 1936), p. 475.

Bibliography of Principal Works Consulted

Abbott, Anthony S. *Shaw and Christianity.* New York: The Seabury Press, 1965.

Barr, Alan P. *Victorian Stage Pulpiteer: Bernard Shaw's Crusade.* Athens: University of Georgia Press, 1973.

The Benedictines of Stanbrook. *In a Great Tradition.* New York: Harper and Brothers, 1956.

Bentley, Eric. *Bernard Shaw.* New York: New Directions, 1957.

Bergson, Henri. *Creative Evolution.* Translated by Arthur Mitchell. New York: The Modern Library, Random House.

Berst, Charles A. *Bernard Shaw and the Art of the Drama.* Urbana: University of Illinois Press, 1973.

Butler, Samuel. *Luck or Cunnng?* The Shrewsbury edition of the works of Samuel Butler, edited by Henry Festing Jones and A. T. Bartholomew, vol. 8. London: J. Cape, 1923.

Carpenter, Charles. *Bernard Shaw and the Art of Destroying Ideals.* Madison: University of Wisconsin Press, 1969.

Chappelow, Allan. *Shaw, "the Chucker-Out,"* London: Allen and Unwin, 1969.

Chesterton, G. K. *George Bernard Shaw.* London: Bodley Head, 1909.

Crompton, Louis, *Shaw the Dramatist.* Lincoln: University of Nebraska Press, 1969.

Darwin, Charles, *The Origin of Species and The Descent of Man.* New York: The Modern Library, Random House, 1971.

Dervin, Daniel. *Bernard Shaw, a Psychological Study.* Lewisburg, PA: Bucknell University Press, 1975.

Dietrich, R. F. *Portrait of the Artist as a Young Superman: A Study of Shaw's Novels.* Gainesville: University of Florida Press, 1969.

Dutli, Alfred. *Der Kosmos eines Ketzers: Die religiöse Bedeutung des Evolutionsgedankes bei Bernard Shaw.* Zurich: Artemis, 1950.

Ervine, St. John. *Bernard Shaw, His Life, Work, and Friends.* New York: William Morrow and Co., 1956.

Evans, T. F., ed. *Shaw, the Critical Heritage.* Boston: Routledge and Kegan Paul, 1976.

Furlong, William B. *Shaw and Chesterton, the Metaphysical Jesters.* University Park: The Pennsylvania State University Press, 1970.

Gould, Stephen Jay. *Ever Since Darwin.* New York: W. W. Norton, 1977.

Henderson, Archibald. *George Bernard Shaw, Man of the Century.* New York: Appleton-Century-Crofts, 1956.
———. *Table-Talk of G.B.S.* New York: Harper, 1925.
Hugo, Leon. *Bernard Shaw, Playwright and Preacher.* London: Methuen and Co., 1970.
Huxley, Julian. *Essays of a Biologist.* Hammondsworth: Penguin Books, 1939. (First published, 1923).
———. *Essays of a Humanist.* New York: Harper and Row, 1964.
———. *Evolution in Action.* New York: Signet, 1953.
———. *Man Stands Alone.* New York: Harper, 1927.
———. *Memories.* New York: Harper and Row, 1970.
Joad, C. E. M. *Shaw.* London: Gollancz, 1949.
Koestler, Arthur. *The Case of the Midwife Toad.* New York: Random House, 1971.
———. *The Ghost in the Machine.* London: Hutchinson and Co., 1967.
Kronenberger, Louis, ed. *George Bernard Shaw: A Critical Survey.* Cleveland: World, 1953.
Lamarck, J. B. *Zoological Philosophy.* Translated by Hugh Elliot. New York and London: Hafner Publishing Co., 1963.
Mander, Raymond and Joe Mitchenson. *Theatrical Companion to Shaw.* New York: Pitman Publishing Co., 1955.
Meisel, Martin. *Shaw and the Nineteenth-Century Theatre.* Princeton, NJ: Princeton University Press, 1963.
Monod, Jacques. *Chance and Necessity, an Essay on the Natural Philosophy of Modern Biology.* New York: Alfred A. Knopf, 1971.
Morgan, Margery M. *The Shavian Playground.* London: Methuen and Co., 1972.
Nethercot, Arthur. *Men and Supermen.* New York: Benjamin Bloom, 1966.
Ohmann, Richard M. *Shaw, the Style and the Man.* Middleton, CT: Wesleyan University Press, 1962.
Patch, Blanche. *Thirty Years With Bernard Shaw.* New York: Dodd, Mead, 1951.
Pearson, Hesketh. *G.B.S., A Full-Length Portrait.* New York: Harper and Brothers, 1942.
Rosset, Ben. *Shaw of Dublin.* University Park: The Pennsylvania State University Press, 1964.
SHAW, The Annual of Bernard Shaw Studies. Stanley Weintraub, general editor. Volume I, "Shaw and Religion," edited by Charles A. Berst. University Park: The Pennsylvania State University Press, 1981.
Shaw, Bernard. *The Black Girl in Search of God and Some Lesser Tales.* Standard Edition. London: Constable, 1934.
———. *Collected Letters, 1874–1897* (Vol. I), *1898–1910* (Vol. II). Edited by Dan H. Laurence. New York: Dodd, Mead, 1965–1972.
———. *Collected Plays With Their Prefaces.* Definitive Edition, edited by Dan H. Laurence. 7 volumes. New York: Dodd, Mead, 1970–1974.
———. *Doctors' Delusions, Crude Criminology, and Sham Education.* Standard Edition. London: Constable, 1932.
———. *Essays in Fabian Socialism.* Standard Edition. London: Constable, 1932.
———. *Everybody's Political What's What?* Standard Edition. London: Constable, 1944.
———. *Immaturity* (a novel). Standard Edition. London: Constable, 1931. (The novel was written in 1879; the autobiographical preface in 1921.)
———. *The Intelligent Woman's Guide to Socialism and Capitalism.* Standard Edition. London: Constable, 1932.
———. *London Music in 1888–89 as Heard by Corno di Bassetto.* Standard Edition. London: Constable, 1937. (Autobiographical preface written in 1935.)
———. *Major Critical Essays.* Standard Edition. London: Constable, 1932.
———. *Our Theatres in the Nineties.* 3 vols. Standard Edition. London: Constable, 1932.

———. *Platform and Pulpit.* Edited by Dan H. Laurence. New York: Hill and Wang, 1961.
———. *Practical Politics.* Edited by Lloyd Hubenka. Lincoln: University of Nebraska Press, 1976.
———. *The Religious Speeches of Bernard Shaw.* Edited by Warren Sylvester Smith. University Park: The Pennsylvania State University Press, 1963.
———. *The Road to Equality.* Edited by Louis Crompton. Boston: Beacon Press, 1971.
———. *Shaw, an Autobiography.* Edited by Stanley Weintraub. 2 volumes. New York: Weybright and Talley, 1969, 1970.
———. *Shaw on Religion.* Edited by Warren Sylvester Smith. New York: Dodd, Mead, 1967.
———. *Sixteen Self Sketches.* Standard Edition. London: Constable, 1949.
———. *What I Really Wrote About the War.* New York: Brentano's, 1932.
St. John, Christopher, ed. *Ellen Terry and Bernard Shaw, a Correspondence.* New York: G. P. Putnam's Sons, 1932.
Smith, Warren Sylvester. *The London Heretics, 1870–1914.* New York: Dodd, Mead, 1967.
Strauss, E. *Bernard Shaw: Art and Socialism.* London: Gollancz, 1942.
Teilhard de Chardin, Pierre. *Christianity and Evolution.* Translated by Rene Hague. New York: Harcourt Brace Jovanovich, 1969.
———. *The Future of Man.* Translated by Norman Denny. New York: Harper and Row, 1964. (A posthumous collection of essays).
———. *The Phenomenon of Man.* Translated by Bernard Wall. New York: Harper and Row, 1959.
Turco, Alfred, Jr. *Shaw's Moral Vision.* Ithaca: Cornell University Press, 1976.
Valency, Maurice. *The Cart and the Trumpet.* New York: Oxford University Press, 1973.
Weintraub, Stanley. *Journey to Heartbreak.* New York: Weybright and Talley, 1971.
———. *Saint Joan Fifty Years After, 1923/24–1973/74.* Baton Rouge: Louisiana State Universtiy Press, 1973.
Weintraub, Stanley, ed. *The Savoy, Nineties Experiment.* University Park: The Pennsylvania State University Press, 1966.
Whitman, Robert F. *Shaw and the Play of Ideas.* Ithaca: Cornell University Press, 1977.
Winsten, Stephen. *Days with Bernard Shaw.* New York: Vanguard Press, 1949.
———. *Jesting Apostle, the Life of Bernard Shaw.* London: Hutchinson, 1956.
Wisenthal, J. L. *The Marriage of Contraries: Bernard Shaw's Middle Plays.* Cambridge: Harvard University Press, 1974.

Index

Page numbers printed in italics indicate the principal discussion of a subject.